Building a Meal

ARTS & TRADITIONS OF THE TABLE

Hervé This

BUILDING

TRANSLATED BY *Malcolm DeBevoise*

A MEAL

FROM MOLECULAR GASTRONOMY TO
CULINARY CONSTRUCTIVISM

COLUMBIA UNIVERSITY PRESS · NEW YORK

Columbia University Press

Publishers Since 1893
New York   Chichester, West Sussex

Construisons un repas copyright © Odile Jacob, 2007
Translation copyright © 2009 Columbia University Press
All rights reserved

Library of Congress Cataloging-in-Publication Data

This, Hervé.
[Construisons un repas. English]
Building a meal : from molecular gastronomy to culinary arts and
traditions of the table / Hervé This ; translated by M. B. DeBevoise
p. cm. — (Arts and traditions of the table : perspectives on culinary history)
Includes index.
ISBN 978-0-231-14466-7 (hard cover : alk. paper) — ISBN 978-0-231-51353-1 (e-book)
1. Cookery. 2. Gastronomy. 3. Food habits—France. 4. Cookery, French.
I. Title II. Series.
TX651.T36813 2009
641.5—DC22
20080406268

Printed in the United States of America

c 10 9 8 7 6 5 4 3 2

References to Internet Web sites (URLs) were accurate at the time of writing.
Neither the author nor Columbia University Press is responsible for URLs that
may have expired or changed since the manuscript was prepared.

BOOK DESIGN BY VIN DANG · JACKET DESIGN BY ISAAC TOBIN

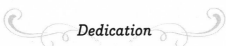

Dedication

It is common for authors to dedicate their books to people whom they love or admire. But what if books were to be dedicated to their readers?

To you, my friend, I dedicate this book, with only one regret, that today's science is imperfect—as I am well aware.

Hurry, it's time to get back to work!

Contents

Desserts

Preface

Eating, cooking, doing science—why are these things fun for some of us but boring for others? And do we *really* love doing what we say we love to do?

Consider eating. Is it actually enjoyable? It involves using some sort of tool (fingers, a fork, sticks) to put pieces of plant or animal tissue into one's mouth, and then activating one's jaw muscles in order to break up the pieces so they are small enough to be swallowed. What's fun about this?

Cooking is hardly more exciting. You take some food products, wash them in water, then toss them into a pan and heat them up. None of this seems very difficult, and in any case there are always recipes to tell us what to do.

And science? You select a physical phenomenon you want to understand, take some measurements in order to characterize it precisely, and organize the data in a series of mathematical relationships (generally linear relationships, since otherwise you won't be able to solve your equations). On the basis of these regularities you propose causal mechanisms, or theories, to explain the phenomenon. You know these theories are probably wrong, but they can be refined and improved by means of prediction and experiment, which in turn generate new data, new regularities, and new theories. The scientist is rather like Penelope in the *Odyssey*, endlessly weaving her tapestry as she waits for Ulysses to come home from Troy.

It's quite true that eating, cooking, and doing science can be boring when they are done mechanically, without any thought or reflection. But think how wonderful they can be when we bring to them a bit of the intelligence that makes us human. Imagine having to cross a bridge every day on the way to and from work. Twice a day, maybe more, we walk from one end to the other, always in a hurry, never paying attention to the bridge itself; indeed, we don't really see it—it's boring, even a nuisance, because it stands between us and our destination. Then one day we bump into an old friend, an architect, and as we are crossing the bridge together she suddenly stops and says, "Look at this tiny bolt! This is what holds the entire bridge together!" Our friend shows us the parts that it joins, and explains that these parts put stress on other parts, just like in the vaults of medieval cathedrals. We stand there for a while, thinking about something whose existence we had never previously suspected, and then we go on our way. Never again will we cross the bridge without admiring the small, apparently insignificant bolt.

Boredom is the result not of doing the same thing over and over again, but of paying no attention to what we are doing. It's no different with eating: If we are content simply to swallow our food without giving it a second thought, it becomes a meaningless chore. Or with cooking: If we approach it only as a technical exercise, it gives us no pleasure. Or with doing science: We have to introduce some sort of spark to get the fire started!

There are so many possible sparks. Conviviality is one, in the case of eating; what is more, there are sound biological reasons for this behavior, as we shall see. In the case of cooking, art is important, but perhaps not more important than the fact that we are human—as we shall also see. And as for science, well, the sort of science I like unites the head with the hand. It is experimental first (though obviously experiments are founded on theory) and formal second (science, Galileo said rather provocatively, is calculation). This, by the way, is why I like certain aspects of *campagnonnage,* a medieval society of craftsmen and artisans that is still very much alive today in France, and that requires apprentices to produce a masterpiece as a condition of membership. In the same vein, I dream one day of having the final say in the matter of bouillon, for example. Why? Because bouillon is the soul of cooking; with a good bouillon, you can make soups, consommés, sauces, and stocks. One day I hope to be able to give a rational explanation for what makes a good bouillon.

This book grew out of conversations with my friend Marie-Odile Monchicourt, a science journalist, portions of which I have reproduced here in order to strike a more personal and informal tone than would otherwise be possible. My purpose is to examine the various elements of our culinary heritage (recipes, principles, adages, time-honored practices, acquired skills, tricks of the cook's trade, dictums, maxims, and so on), to scrutinize familiar assumptions, observe phenomena, and search for the mechanisms that underlie them. My aim is not to destroy or deconstruct our traditional ideas about cooking, but rather to renew a heritage: the cooking that we have inherited from the past and that has gone unquestioned for centuries. What should we conserve from the past? What can we do without in the future? What can we transform and improve? Science can help us answer these questions.

Science never ceases to arouse our curiosity, to nourish our minds with fresh insights, and to make us look at nature and the world around us in new ways. As the universal language of conviviality, it can guide us in exploring cooking and changing the way we think about food—a revolution in which every one of us can take part!

The following pages contain many references to my friend Pierre Gagnaire. I ask the forgiveness of all the many other culinary artists whom I admire but whom I mention seldom or not at all, among them Michel Bras, Olivier Roellinger, Philippe Conticini, Christian Conticini, Pascal Barbot, Heston Blumenthal, Michel Roth, Ferran Adrià, Andoni Luis Aduriz, Alain Llorca, François Pasteau, Alex Atala, and Georges Roux. All of them, by their work and through their occasional writings, contribute to the advancement of culinary art; their achievements adorn the great monument of cuisine they are helping to build. Pierre happens to be a close friend, the person with whom I most often share ideas about cooking, but quite obviously my other friends and acquaintances are not excluded from our discussions. We seek consensus, not controversy. Everyone is invited to the grand feast of knowledge.

Building a Meal

Introduction
CULINARY CONSTRUCTION

*When there will no longer be cuisine in the world, there will no longer
be letters, elevated and lively intelligence, or sociability. There will no
longer be social unity.*
MARIE-ANTOINE CARÊME (1783–1833)

To construct a meal, isn't it enough to put together selected dishes from the classical repertoire? Yes, but why should we go on doing what has already been done before? In this book I will not go so far as to try to completely reinvent cooking. I will halt at the point where constructing a meal makes it possible to ask questions. Questions, marvelous questions! Don't they deserve answers—invitations to go further, instead of contenting ourselves with what we already have? Aren't questions the seeds of discovery and invention?

In the course of creating a menu we will make a few detours, because idling along the way is also a form of discovery, as the word *method* says (it comes from the Greek *methodos*, "the way of pursuing knowledge"). Attracted by the most varied riches of existence, the mind adorns itself with a thousand memories—memories that, one day, will crown its journey. And in any case, isn't the journey more important than the destination? Whoever climbs a mountain with the thought only of getting to the top is unhappy the whole way up, says Eastern philosophy; whereas the person who takes pleasure in every step proceeds slowly, but in a state of happiness that leads him imperceptibly to his destination—a destination that is marvelous, because inaccessible. The nineteenth-century French chemist

Michel-Eugène Chevreul (1786–1889), who in later years liked to refer to himself as France's "oldest student," took as his motto, Strive for perfection without pretending to have achieved it. Is this Gallic wisdom? Buddhist wisdom? Universal wisdom?

The Complexity of the World

To cook is to use ingredients, which is to say complex physicochemical objects. Is their behavior amenable to scientific investigation? Considering that the mathematician Henri Poincaré (1854–1912) demonstrated that the trajectory of only three bodies interacting under the force of gravity is not calculable, one may well wonder.

Foods are complex, yes; but not excessively so, as long as we have a suitable method of analysis at our disposal. Let me stop right here and point out that this book is first and foremost a book of methods, and only secondarily one of facts. We will, of course, end up cooking the meal referred to in the book's title, but we will often have occasion to discuss methods.

In thinking about ingredients, we are frequently faced with the need to characterize them by means of a first approximation, then a second approximation, and so on. What do I mean by this? It is useful, for example, to recognize that tea is mainly water (not completely, of course, for otherwise tea lovers would not go to the trouble of infusing for exactly three minutes, carefully choosing their water, their tea pot, their cups). But tea—like coffee, bouillon, wine, and the juices of meats—is water only at a first approximation. At a second approximation, or level of analysis, these liquids contain various molecules, much less numerous than water molecules, that nonetheless impart a particular flavor that we appreciate. Since we are talking about cooking, it is useful to distinguish between a primary chemical level of analysis and a primary gustatory level. The former is what we detect when we consider the component elements of a food. At a first level of chemical analysis, tea is water, oil is fat, sugar is sucrose, flour is starch, and so on. But from the point of view of flavor, tea is not water at a first approximation (indeed, the water is secondary); nor is olive oil to be confused with pistachio oil or hazelnut oil. At a first level of flavor, water consists chiefly of ions, which give water its flavor, contrary to what we are taught at school (namely, that water is odorless, colorless, and tasteless); tea, for its part, owes its organoleptic properties to a variety of marvelous molecules, among them phenolic compounds.

This distinction between a first *chemical* approximation and a first *flavor* approximation is essential, because we eat foods, not chemistry. The flavor of coffee is not the same as that of tea; chocolate—fat and sugar, at a first level of chemical analysis—is not a mixture of fats and sugar from the point of view of flavor. These sorts of distinctions, between chemistry and flavor, art and craft, innovation and invention, and so on, illustrate the kinds of methods that will accompany us on our voyage toward the construction of a meal.

The menu? To begin with, we need to keep in mind that small portions make for a more refined cuisine, and more discriminating diners, than heaping portions (why this should be so will need to be examined, by the way). But that's no reason for us to starve ourselves. Let's bring our appetite and be prepared to sit down to a meal with two appetizers, two main courses, and two desserts—larger than some of the meals served by great culinary artists.

For our exercise in constructing a meal, we will stay in the realm of classical cooking, except for the final dessert, for reasons that I will explain along the way and that have to do with a peculiar characteristic of the human race known as food neophobia.

Here, then, is the menu I propose:

APPETIZERS

Hard-boiled egg with mayonnaise
Simple consommé

MAIN COURSES

Leg of lamb with green beans
Steak and French fries

DESSERTS

Lemon meringue pie
Futuristic chocolate mousse

Each recipe will be the occasion for specific digressions, anecdotes, and so on, and at the end of each one we will examine the path that cooking must follow in the future if it is to improve. Yes, even the humble hard-boiled egg with mayonnaise will enable us to win a third star, if we set ourselves the task of making the perfect hard-boiled egg.

I

Hard-Boiled Egg with Mayonnaise

To make a traditional hard-boiled egg with mayonnaise, we have to hard-boil an egg, on the one hand, and make a mayonnaise sauce on the other. Immediately, two questions arise: Why should we want to make this dish in the first place, and why should we insist on this particular combination of an egg and a sauce?

Why the familiar hard-boiled egg with mayonnaise? Lurking behind this question is another one, having to do with tradition. Etymologically, tradition is that which is handed down. But if we have been exposed at a very young age to new methods, what is traditional for others is obsolete for us, and what is traditional for us seems innovative to our neighbors. We need to be wary of doing something over and over again, simply because it is traditional. But what of something that is "classic"? Here the sense seems better suited to our purposes. Whereas in 1548 the word in French meant "that which is among the best things," already by 1611 it had come to mean "that which has authority, that which is considered a model of its kind"—the sense that it still retains today.

Let's set ourselves the task, then, of making not a traditional hard-boiled egg, but instead a classic hard-boiled egg. The standard works to consult in this connection are *La Suite des dons de Comus* (1742), by François Marin, *L'Art de la cuisine française au XIXe siècle* (1854), by Marie-Antoine Carême, and the *Dictionnaire universel de cuisine* (1889–92), by Joseph Favre. These authors give two sorts of recipes. The first recommends immersing the egg in boiling water for ten minutes; the second calls for the egg to be placed in cold water that is then heated.

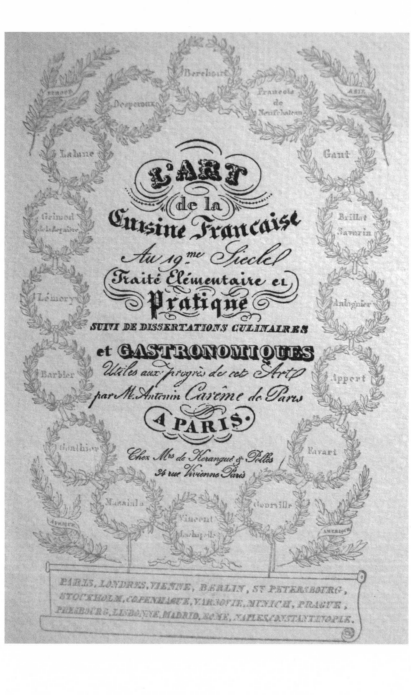

There has been much disagreement among professional cooks over the humble hard-boiled egg, but common sense shows that the second method is no good. If you start with cold water, what will happen to the egg as it is heated? If the water is heated very slowly, the egg will spend a good deal of time at 20°C (68°F), then at 30°C (86°F), 40°C (104°), 50°C (122°F), and so on, until the water finally comes to a boil; notice that it will pass a very long time between 90°C (194°F) and 100°C (212°F), eventually bringing the internal temperature of the egg to the boiling point. A simple experiment will show that hard-boiled eggs made in this way are dreadful. By contrast, if one puts the egg in boiling water, it is immediately heated to 100°C and remains at this precise temperature for the entire time that it is in the water. The cooking can therefore be regulated more precisely.

Heresy!

Do we have the right to challenge culinary tradition? Must we "respect" the culinary tradition in which we were brought up? Do we dare assume the responsibility of helping cooking to evolve?

Molecular gastronomy, the subject of this book for the most part, is a scientific discipline. In other words, it is a field of research; it is neither a method of instruction, nor a technology, nor a technique. What is the difference between these things? Technique is the execution of deliberate movements—which have their origin in the mind, of course. Instruction involves the transmission of knowledge. Technology is a matter of improving technique and applying new knowledge produced by science, of transferring knowledge from the realm of science to that of technique. Science, for its part, is a way of exploring the world.

Thus it is that cosmology explores the universe, that geology seeks to understand the terrestrial world, botany to understand plants, zoology to understand animals, chemistry to understand chemical reactions, physics to understand physical phenomena (optics, acoustics, mechanics, and so on). Each science has it own object of study, all of them parts of the world in which we live.

The first three volumes of Marie-Antoine Carême's monumental five-volume work *L'Art de la Cuisine Française* appeared in 1833. Completed by Carême's student Armand Plumery, it remains the standard reference on classic French cuisine.

Molecular gastronomy studies the part of the world occupied by cooking and, to some extent at least, degustation—a very significant part, it must be said, considering that every home, no matter how small, has a kitchen. A branch of science therefore had to be developed to study its mysteries. Can we say more precisely what science is? Science is the activity of observing phenomena, searching for mechanisms and explanations, and proposing models and theories. In the case of molecular gastronomy, it seeks to identify the mechanisms of culinary transformations, which are essentially chemical, physical, or biological in nature. The method of study employed by molecular gastronomy—here again, and everywhere in what follows, we encounter this notion of method—is the same method that all sciences use, the experimental method, envisioned by such noble minds as Aristotle (384–322 BCE), Roger Bacon (c. 1214–94), Francis Bacon (1561–1626), and Galileo (1564–1642).

The experimental method proceeds by examining a physical phenomenon that has been identified. To be able to study the blue of the sky, you must first have seen the blue. In order to observe that eggs, which are liquid in their raw state, solidify when they are placed for several minutes in boiling water, you must leave them in the water long enough for coagulation to take place. Then various measurements can be made to characterize the phenomenon, determining the temperature inside the egg, the degree to which the egg hardens, and so on. You need to be sure that all the data collected are relevant. What would be the use of measuring the ambient temperature in the kitchen when cooking an egg in boiling water?

Because phenomena occur over time, parameters such as water temperature vary between the initial state (when an egg is placed in the pan) and the final state (when the water has reached 100°C [212°F]). This makes it necessary to take measurements at regular intervals and then analyze the recorded variations mathematically by means of differential equations. With the aid of physical laws we can use these experimental results to discover a great many interesting things. Thus, for example, we can determine the minimum amount of energy needed to boil a particular egg by multiplying the mass of the egg by its thermal capacity and the difference in temperature between the initial and final states. Laws condense an infinite number of possibilities. In the case of boiling eggs, the same law, or formula, gives the energy for eggs of any mass.

On the basis of measurements and the relevant laws, we then search for a theory (for example, that the egg whites harden because their proteins are joined together in a network that traps the water molecules). Since a theory is a sort of simplified model of reality, and since we know perfectly well that simplified models are not to be confused with the objects they represent, we then seek to disprove the theory (even—or especially—if it is our own). This refutation may be attempted by calculation (it is very easy, for example, to calculate that the quantity of proteins present in an egg as it hardens is far greater than the quantity of proteins needed to make an aspic set, which enables us to make predictions about the hardness of the hardened egg) or by experiment.

In the event that the attempt at refutation has succeeded, we either abandon or modify the theory (instead of asserting that the egg white is 90 percent water and 10 percent proteins, for example, we say that 90 percent of it is water and the rest is made up of various proteins, all of which have different coagulation properties). Then we try our best to refute this new theory, refining it further as necessary, and so on.

Unlike pseudoscience, which typically countenances no challenge to its claims, genuine science is never satisfied with itself. It recognizes that every theory is false or, at least, that every theory describes phenomena imperfectly. And because it is wary of certainty, regarding even what is probable as false until evidence to the contrary can be produced, science avoids dogmatism, smugness, and overconfidence in its own powers.

Renewing a Heritage

The question of tradition vs. classicism can be posed in more specific terms. If we learn to make airier soufflés, mayonnaises with new and different flavors, lighter mousses, gnocchi that are more evenly cooked, chocolate mousses without eggs, and so on, we run the risk of changing the cuisine we've grown up with.

IS THIS REALLY A GOOD IDEA?

I suggest that we think of classical French cuisine—*la grande cuisine française,* of which the French are rightly so proud—as an old ancestral home. It is a superb

Culinary constructivism is a new movement that aims to produce dishes without reference to those of the past, taking into account only their gustatory effects. Shown here is a concoction by Pierre Gagnaire.

building, but utterly lacking in modern comforts: no indoor bathrooms, toilets out in the back, an old wood-burning stove in the kitchen. Should we keep the house as it is, despite its many inconveniences? Selling it or tearing it down would be irresponsible, because all the intelligence it embodies (much of which we scarcely stop to consider) would be lost to us. Its thick stone walls do a better job of protecting us against cold in the winter, and against heat in the summer, than modern asbestos-coated cinderblocks. No, we would be wise to keep the house, while at the same time equipping it with modern appliances so that we may live in it more comfortably than our ancestors ever did. Owing to a lack of hygiene, among many other things, their life expectancy was far shorter than ours; their wine often turned sour; their fruits were frequently blemished, their eggs less fresh, and so on.

Renovation is clearly in order, then. But how to are we to go about it so that we won't regret clumsy alterations later? The question cannot be avoided, because ill-considered changes can have terrible consequences.

The Birth of Molecular Gastronomy

Marie-Odile Monchicourt: How was molecular gastronomy born?

Hervé This: In 1988 a Scotswoman who had worked in the editorial department of *Europhysics Letters*, the journal of European physicists, joined the staff of *Pour la Science* (the French edition of *Scientific American*), where I worked for twenty years. On learning that I was fascinated by cooking and was exploring its physical chemistry in a home laboratory in my spare time, she told me about a physicist at Oxford named Nicholas Kurti, a former president of the Royal Society, who was conducting research similar to mine. I telephoned Kurti and some days later he came to see me in Paris. It was the beginning of a wonderful friendship and of an adventure that was to lead to molecular gastronomy.

M.-O. M.: Was Kurti doing the same work you were?

H. T.: Not quite the same, no. I was chiefly interested in traditional practices, adages, dictums, tricks of the kitchen, and so on, whereas he wanted to modernize cooking by introducing experimental methods developed by physicists. Our first meeting took place at a restaurant in Paris, Chez Maître Paul, on the rue Monsieur-le-Prince, where we shared a chicken in a straw wine and morel sauce. Kurti was then eighty years old, a renowned physicist known particularly for devising a cooling method known as nuclear adiabatic demagnetization, thanks to which he had been able to achieve the lowest temperatures ever, a few fractions of a degree above absolute zero (-273.15°C [-459.67°F]).

We immediately became friends and communicated with each other almost every day, comparing experimental results, writing articles together, and giving joint lectures. We soon conceived the idea of holding an international symposium at Erice, in Sicily. We needed a name for what we were doing and for the symposium itself. I proposed "molecular gastronomy," but Kurti, as a physicist, feared that this assigned too much importance to chemistry (since some culinary transformations can be explained macroscopically), so we finally agreed on "molecular and physical gastronomy."

The first international conference of the discipline took place in 1992, bringing together chefs and scientists from around the world in a convention center that normally hosted particle physics meetings. Three years later, we organized another international conference that was attended by a Nobel Prize–winning physicist, Pierre-Gilles de Gennes. At the time I was defending my doctoral thesis, likewise entitled "Molecular and Physical Gastronomy," before a jury that included de Gennes and Jean-Marie Lehn, a Nobel laureate in chemistry who had invited me to conduct my research in his laboratory for the chemistry of molecular interactions at the Collège de France in Paris.

M.O.-M.: Thereafter, I gather, the two fathers of molecular gastronomy continued to devise new research projects as well as new ways of disseminating the results of this research.

H. T.: When Nicholas Kurti died, at the age of ninety in 1998, I renamed the discipline simply "molecular gastronomy," while baptizing the international conferences in Sicily, which continue to take place every two years or so, with Nicholas's name as a permanent memorial.

Return to the Kitchen

The point of the metaphor of the ancestral home is readily grasped. In much the same way, by questioning the foundations of classical cuisine, we risk pulling down a grand edifice. No one disputes that the building needs to be renovated—but which parts of it? What should we preserve, and what can we safely consign to oblivion? No less than in the case of an old house, the responsibility to the past forms a crucial part of the work of the scientist. In the words of Rabelais, "Science without conscience is but the ruin of the soul."

The curious thing is that in the realm of cooking the question of preservation should be posed by scientists and not by cooks themselves, who have blithely gone about changing it in various ways, following their own aesthetic tastes. No one today any longer makes custard, for example, the way people did a hundred years ago. The number of egg yolks per quart (as many as sixteen) seemed excessive, and it was reduced without anyone wondering whether there was a law against changing the proportion. Cooks fixed up this or that room of the ancestral home without trying to form an overall idea of it, without imagining the long-term consequences of what they were doing.

The time has finally come to ask what we can renovate and what we ought to preserve. Posing these questions systematically, institutionally, experimentally, and analytically is the only way that the necessary transformations can be responsibly carried out.

How to Make a "Good" Hard-Boiled Egg

Since a general perspective or overview of the whole of cooking is beyond the reach of most of us, we will find it much more useful to content ourselves for the moment with taking a narrow view of the matter, if only because this is the view that we are used to taking.

Let us therefore come back to our hard-boiled egg. We have to cook it at some point anyway, and we want it to turn out well. Indeed, we would like it to be better than good—perfect, if such a thing is possible in this world. This is all very well, but what does a perfect hard-boiled egg look like? Let's make a list of the properties of the egg of our dreams:

> *its shell must not crack during cooking;*
> *it must easily peel off from the egg once cooked;*
> *the white must not be rubbery;*
> *the yolk must not be sandy;*
> *the egg must not smell of "sulfur" and the yolk must not be greenish;*
> *the yolk must be perfectly centered in relation to the albumen, so that when sliced the*
>   *egg exhibits a pleasing symmetry.*

These, then, are the properties of an ideal hard-boiled egg ... for me. Let's examine them in turn.

THE EGGSHELL MUST NOT CRACK DURING COOKING

To prevent the shell from cracking it is usually recommended that vinegar be added to the cooking water, or salt, or even burned matches. Some cookbooks reject this advice, however, emphasizing instead the need to avoid the "thermal shock" that occurs when the egg is put in boiling water. I invite you to join me in testing both of these culinary dictums (as I call such old wives' tales, cooking tips, adages, pieces of homespun wisdom, tricks of the cook's trade, and other anecdotal bits of cleverness)—in order to satisfy yourself that neither of them is any good.

Culinary Dictums by the Thousands

M.-O. M.: True dictums, false dictums—why do you attach such importance to them?

H. T.: First, because a dictum concerning the preparation of a Roquefort soufflé lies at the root of molecular gastronomy, which is now developing throughout the world with the creation of seminars, national associations, university chairs, conferences, training programs, and so on.

I have had a lab at home since I was six years old. When I was a student at the École Supérieure de Physique et de Chimies Industrielles in Paris, I lived in a small room where my pots and pans vied for space with test tubes. I had gotten rid of all my chemicals because I had everything I needed at the university lab. All my scientific equipment at home had been diverted to culinary purposes.

On 16 March 1980 I had my *nuit de Pascal*—the experience that changed my life. It was a Sunday evening and I'd invited some friends over to dinner. Using a recipe for a Roquefort soufflé, I came across a culinary dictum: After making the cheese béchamel sauce, you "add the yolks two by two." Why two by two? Since there didn't seem to be any particular reason to do this, I put in all the egg yolks at once—and the soufflé turned out poorly. I didn't know why.

The following Sunday some other friends were coming for dinner, so I figured I might as well try the soufflé again. Coming across the same phrase—add the yolks two by two—I wondered whether the failure of the first soufflé was due to putting all the yolks together. But if adding the yolks two by two yielded a better result, why wouldn't adding them one by one give a still better soufflé? So that's what I did, and the result was in fact better—or so it seemed. (We know now that adding the yolks one by one, or two by two, or all together doesn't affect the success of a soufflé.) At any rate, this event was the beginning of a collection of culinary dictums that I am still adding to today.

M.-O. M.: How many of them do you have?

H. T.: Since 1980 I've collected more than twenty-five thousand such dictums, mainly from cookbooks in French. Each culinary dictum needs to be rigorously tested. This testing has made up a large part of the work of molecular gastronomy, which, as I say, was founded eight years later, in 1988.

One finds dictums of all sorts in cookbooks, ancient and modern alike:

- *ones that seem to be good and that in fact are good;*
- *ones that seem good but aren't;*
- *ones that seem bad but that in fact are good;*
- *ones that seem bad and indeed are bad;*
- *ones whose correctness or incorrectness cannot be determined simply by logical deduction;*
- *and ones whose status has changed over the centuries, because foods and cooking techniques have changed.*

Here are a few examples:

- *When making a bouillon, the pot must be partly covered, leaving an opening two fingers' wide.*
- *For a soufflé to rise just as it should, the egg whites must be beaten until they are quite firm.*
- *A spoonful of boiling vinegar must be added to a mayonnaise sauce for it to hold together.*
- *Green beans are greener if they are cooked in an uncovered pot.*
- *Squid are more tender when they are cooked in water containing burned matches.*
- *The odor of cauliflower is eliminated by immersing a bread crust in the cooking water.*

And so on. If we assume that rigorously testing a dictum requires between a week and three years of work—on average, let's say, a month—then verifying all the culinary dictums that I have collected would take twenty-five thousand months, or about two thousand years—longer than anyone's lifetime, in any case. And so I asked universities and professional cooking schools in France to help out. Assuming that professors and students at five hundred institutions were to test five dictums per year, it would take ten years to gather the necessary data—though not to analyze them. The empirical part of the enterprise has now begun in a number of French culinary programs under the aegis of what are called molecular gastronomy workshops.

Not to be outdone, scientists and cooks in other countries—including Greece, Argentina, Switzerland, the United States, Denmark, Cuba, Brazil, Portugal—are now collecting culinary dictums found in their own culinary cultures and testing them.

It is true that vinegar facilitates the coagulation of the albumen that seeps out of a cracked egg; at the same time it weakens the shell by dissolving it. It is also true that salt gives flavor to the egg white by diffusing through the pores of the shell, and it certainly helps to avoid the phenomenon of osmosis, which causes the liquid inside the egg to swell and increases the chance the shell will crack.

Nonetheless, theory needs to be tested by experiment. For example, if you cook a large number of eggs in salted water, and an identical number of eggs in pure water, you will see that the number of cracked shells is comparable in both cases. Further tests will show us that many such received notions must be revised, and lead us to ask why they were accepted without question. Rather than blindly apply the culinary dictums accumulated from past experience, rather than recapitulate the empirical history of cooking in the hope of identifying an effective method, why don't we use reason instead?

Indeed, reason is the only effective method. In this case it will lead us to the simple conclusion that the shell of the egg encloses not only the white and the yolk but also a pocket of air, especially when the egg is not absolutely fresh. This air pocket becomes larger as the egg gets older, for the water of the egg white evaporates through the pores in the shell—which is why the freshness of an egg can be determined by placing it in heavily salted water. If the egg is not fresh, its large end (where the air pocket is located) floats and keeps the egg near the surface. During the course of cooking, the liquid inside the egg expands a little. Not much (by only one twenty-thousandth of the volume), but a little just the same; and because this liquid is incompressible, the fact that it cannot escape the shell means that in expanding it is bound to exert pressure on the shell. Theoretically, osmosis may play a role, but since the air pocket contains a gas, which, unlike the liquid, is compressible, it can absorb the compression due to the expansion of the liquid.

To help the gas escape—and so prevent pressure from building up inside the shell—you have only to take a pin and make a small hole in the large end of the egg. When the egg is in the boiling water, you will therefore see small air bubbles escaping from the shell—with the result that the egg doesn't break.

THE SHELL MUST PEEL OFF EASILY

How can we make an egg easy to peel? It is hard to remove the shell from some eggs, which means that, when the shell sticks to the membrane that covers the egg and the membrane sticks to the egg white, bits of the albumen come off along with the shell. Some cookbooks recommend putting the eggs in cold water; others say to let the eggs cool before peeling them; still others blame the coolness of the eggs without suggesting any solution to the problem.

Once again, we are left with a choice between reflection and deduction, or the experimental method. I took advantage of a laboratory session with students at the University of Tours to test eggs of all sorts, treated in every way imaginable: taken out of the water to cool, left in the hot water, placed in cold water, and so on. No method proved to be better than any other, unfortunately. It therefore occurred to me that we should look to chemistry for what it does best—giving us knowledge about the properties of materials and the transformations they are capable of undergoing. The shell of an egg is formed almost exclusively of calcium carbonate, which we know reacts with acids to form carbon dioxide (a gas under usual conditions); in the process, the calcium ions of the carbonate are dissolved by the acid. This is the reason, for example, why you should not clean marble with vinegar. The vinegar dissolves the marble instead of removing stains.

In the case of an egg the corrosive effect of vinegar is much more interesting, because the shell softens and then completely disappears. The reaction is very gradual, of course, but one could use stronger acids. And since any acid is apt to make the egg inedibly sour, one could neutralize the surface of the egg afterward with the aid of a gentle alkali, such as baking soda (sometimes used in the kitchen to soften dry vegetables when the water is too hard—but that's another story).

The nonchemist who smiles at the idea deserves to be told that the process of dissolving the shell is no more "chemical" than the ancient Chinese technique for making "hundred-year-old eggs"—eggs preserved in a plaster of straw, mud, lime, and ashes. Chemists know that ash is a source of potash, that is, potassium hydroxide, which is an alkali or a base, like lime, soda, and ammonia. When eggs are in contact with these substances for an extended period, the white coagulates, and a hard egg with a characteristic flavor is gradually formed. Conversely, a fresh egg placed in an acid for several weeks slowly coagulates, after its shell has been dissolved, forming what might be called an "anti-one-hundred-year-old egg" (since acids are the opposite of bases).

The moral of the story is that vinegar attacks the shell. Therefore, to remove the shell without damaging the egg, we can place the hard-boiled egg in vinegar for a few hours.

THE WHITE MUST NOT BE RUBBERY, NOR THE YOLK SANDY

Let's come back to more classical procedures. When you place an egg in boiling water for ten minutes, you obtain a coagulated egg white and a coagulated yolk. Why ten minutes? This length of time does not figure in the oldest cookbooks, but it seems to be something like an empirical decree, for an egg that has cooked for longer than ten minutes has a rubbery white and a sandy, greenish yolk. It is true that some people like their hard-boiled eggs this way, but most people can't stand them. Without taking sides, we should admire the unusual phenomenon that we are dealing with. Yes, coagulation is a strange transformation that causes a yellowish green liquid to turn white as it solidifies. This is not a "natural" transformation, at least not by comparison with the fusion that occurs when water freezes, forming ice, or the melting of a piece of butter. Is it therefore artificial? Certainly, because it takes place only as a result of human activity, in this case the activity of a cook.

Bread is obtained by kneading flour with water, salt, and yeast. What would you get if the water were flavored? Let's try substituting tea, coffee, bouillon, orange juice . . .

Indeed, it is a phenomenon that borders on the miraculous, one that is no less remarkable in its way than the transformation of water into wine. (In saying this, by the way, I do not blaspheme. A believer cannot contemplate such a transformation without praising God, and a nonbeliever cannot fail to be in awe of the stunning transformations of which matter is capable.) How does coagulation occur? Are factors other than temperature at work? And if only temperature matters, how many degrees are necessary to bring about the transformation? In other words, at what temperature is an egg white cooked?

At 100°C (212°F) an egg coagulates, of course, but not at 20°C (68°F). What about 40°C (104°F)? Heat waves answer this question: Hens continue to produce chicks, so evidently their eggs do not cook at this temperature. By the way, the qualification "hen's egg" is not a trivial point. As Jules Verne noted in *L'Île mystérieuse* (1874), sea turtles' eggs do not coagulate, even in boiling water. Why? I don't have the faintest idea! Very well, then, at what temperature is a hen's egg cooked? The short answer is that the white begins to coagulate at 61°C (142°F) and the yolk at 62°C (144°F). But there is more to it than that.

At What Temperature Is an Egg Cooked?

Let's take a step back for a moment in order to appreciate the implication of the previous paragraph more fully. So far I have been circling around the question, soliloquizing in the manner of Denis Diderot. Paul Valéry said that a writer is someone who doesn't find words that satisfy him, and so he goes out looking for better ones and finds them. One might say, then, that a scientist is someone who doesn't understand phenomena, and so she manipulates them (by means of thought experiments and actual laboratory experiments) in order to better understand what is going on. Were Albert Einstein's thought experiments anything other than a way of manipulating phenomena? Einstein discovered the theory of relativity by asking what he would see if he were seated astride a rocket traveling at the speed of light. Would he see light, since he was traveling at the same speed and therefore—because light can't exceed its own speed—couldn't be overtaken by it?

Let us therefore come back to our little question, abandoning all scientific pretension: At what temperature is an egg cooked? As before, two approaches are available to us. Either we conduct an experiment or we critically examine the existing theory of the phenomenon. This time, let's do an experiment. Take a flat-bottomed glass, put some egg whites in it, and then place the glass on a cold electric coil. Now gently heat the egg whites, through the bottom of the glass. Don't forget to put a thermometer in the glass, which will tell us the temperature at the depth to which it is inserted. (The best device for this purpose would be a thermocouple—a sensor composed of two soldered prongs and a microamperemeter, which measures the difference in electrical potential between the two free extremities of the prongs—because it gives the temperature instantaneously, and much more precisely than classical mercury or alcohol thermometers.)

Gradually the temperature rises at the bottom of the glass, which is soon filled by a white coagulated layer. The coagulation boundary slowly rises in its turn, allowing us to place the thermometer (or, far better, the thermocouple) both below and above the boundary. We will find that the temperature of the coagulated layer is between 60°C (140°F) and 70°C (158°F).

Is there any need for it to reach a temperature of 100°C (212°F)? No, but this degree of cooking has the merit of being easily identified, which was very useful in earlier centuries when thermometers were far rarer items than they are today.

What, then, is the exact temperature at which coagulation occurs? To answer this question, we have to devise a more precise experiment. Let's begin by heating some eggs in water. More exactly, let's heat some eggs in a large quantity of water, so that the thermal inertia of the system allows them to remain at successive intervals for a time—long enough to observe what occurs first at 60°C (140°F), then 61°C (142°F), then 62°C (144°F), and so on. After a few minutes at a given degree of temperature, take an egg out from the water and see how far it is cooked.

Doing this we observe that the egg white begins to coagulate at 61°C—and that it continues to coagulate, not as a function of time but of temperature. In other words, whether we heat an egg to 61°C for a minute (note that this must be the temperature throughout the egg) or for an hour, it will always be cooked to the same degree. By contrast, an egg heated to 70°C (158°F) will be different from an egg that has been heated only to 61°C, and an egg heated to more than 80°C (176°F) will be more different still, and so on. Why?

Because an egg white is composed of 90 percent water and 10 percent proteins. Both the water and the proteins are molecules. In the case of the water, we may imagine billiard balls that are always in motion, at a speed that increases with the average temperature; for the proteins, imagine minuscule pearl necklaces dispersed among the water molecules. What assurance do we have that this picture is accurate? Science cannot prove that water consists of molecules. This is because, strictly speaking, science demonstrates nothing; it limits itself to refutation. If it is assumed that water is made up of molecules that behave like billiard balls, science can show that their average speed does not diminish with the temperature. Or it can show that water molecules, if they exist, must be smaller in diameter than the wave length of visible light. But it cannot demonstrate the actual existence of molecules. It can do no more than supply certain indications, which, in the best case, are subsequently corroborated by further experiment.

An egg white is composed of folded packets of proteins dispersed in water.

Take an hourglass, for example, and turn it over. The grains of sand drain out into the lower compartment. With a much finer powder the phenomenon would be the same, but the grains would be imperceptible, and, if there were enough of them, you would have the impression of seeing the flow of a liquid. This sort of analogy, which leads us to regard matter as being granular in nature (we call the grains of which it is composed molecules, which in turn are made up of atoms), is nonetheless inadequate, because it does not explain why a drop of ink carefully placed in a stationary glass of water ends up being dispersed throughout the liquid and, once diluted, disappears from view.

To account for this latter phenomenon, we make the further assumption that the water molecules are like billiard balls that are always in motion, only too small to be seen; they knock into the ink molecules, dispersing them. This hypothetical mechanism provides a plausible explanation of the phenomenon, but it nonetheless does not amount to a formal demonstration.

When egg white is heated, the motion of the protein molecules is accelerated along with that of the water molecules. The protein molecules, as I say, may be thought of as miniature strings of pearls that are folded in on themselves. The shock of their collision with the water molecules causes them to be unfolded, exposing the protein chains that had been tucked away inside to chemical reaction. In this way the unfolded packets become attached to one another, forming a sort of three-dimensional fishing net, the mesh of which is made of proteins. The water molecules are the fish that are caught in the net. This is why even if albumen is made up mostly of water, it undergoes a physical transformation, passing from a liquid to a solid state without losing any water during cooking. The water molecules are still there, but they have been trapped in the protein net (or network, as scientists call it).

Why does coagulation proceed by stages? Because the egg contains more than one sort of protein, and each sort is unfolded (or "denatured") at a particular temperature, which depends in turn on the amino acid residues that characterize its subunit composition.

At 61°C a first sort of protein coagulates, forming a very delicate gel. Once a higher temperature has been reached,

The coagulation of egg white occurs when the proteins combine to form networks that trap the water molecules.

another sort of protein coagulates in its turn. Just as two nets do a better job of trapping fish than one, two coagulated protein networks yield a firmer egg white than only one network. This process continues, with the result that the number of types of protein that coagulate increases—and the egg white hardens.

The yolk? It begins to coagulate at 62°, but because only very weak concentrations of protein are involved at this stage, the transformation is not visible to the naked eye. Coagulation begins to be visible only at 67° to 68°C (153°to 154°F).

THE EGG MUST NOT HAVE A SULFUROUS SMELL AND THE YOLK MUST NOT TURN GREENISH

When an egg is left in boiling water for more than ten minutes, a sulfurous odor can be detected. But this odor does not occur when an egg is left overnight in 70°C (158°F) water. Why should this be?

Many societies have learned to prevent sulfurous odors from forming. The hamine eggs of Jewish communities in Turkey and Greece are cooked in warm ashes for an entire night without giving off the smell of sulfur. The hot springs eggs (*onsen tamago*) found in Japan are similar. In both cases, the eggs are cooked at well under 100°C, and a sulfurous odor is avoided (not always altogether successfully, however, since it is difficult to control the temperature of ashes or hot springs).

Note, too, that the familiar odor associated with sulfur cannot really be due to sulfur, which typically is encountered in the form of a yellow, odorless powder. However this yellow element, when burned, reacts chemically and emits a nauseating smell. One of its compounds in particular, dihydrogen sulfide, is well known to chemists for its smell of rotten egg. This gas is classically detected by filter paper that absorbs a completely colorless lead acetate solution. When the lead acetate is exposed to dihydrogen sulfide, the two substances react and produce a dark precipitate. This precipitation is a better indication of the presence of the sulfur gas, because it is toxic. Place lead acetate paper some centimeters above an egg white or a yolk as it is being cooked, and you will discover that heating to sufficiently high temperatures causes the nauseating gas to appear.

And this same gas, formed inside the egg, reacts with the iron atoms of certain proteins to create the greenish ring found in bad eggs. In other words, the egg's strong odor appears at the same time as the disagreeable color—along with a most indelicate flavor. How can these things be avoided? By heating the eggs to a temperature well below 100°C.

THE YOLK MUST BE WELL CENTERED

Some cookbooks say that in order to center the yolk in a hard-boiled egg, the egg must be cooked in boiling water. Others say that it should be placed in cold water. Which view should we credit? Science textbooks always show the yolk at the center of a raw egg. But if it appears at the center when cooking begins, why shouldn't it be there at the end? Since the yolk is generally found to be decentered after cooking, it would appear that the yolk was not centered in the egg to begin with—and that science textbooks, like cookbooks, are sometimes mistaken.

Bringing our faculty of understanding to bear on the problem, as natural philosophers used to say, let's proceed slowly and rationally. We begin with an egg, which we hold before us, vertically, with the small end pointing upward. Where is the yolk? For reasons of symmetry, there is no reason that the yolk should be anywhere else than along the vertical axis, but the question is at what level: at the bottom of the egg? in the middle? at the top? Interestingly, when this question is put to a lecture audience, most people answer "at the bottom."

Without prejudging the truth or falsity of this answer, we should start by asking what prevents us from seeing the yolk in the egg. The shell, quite obviously, because it is opaque. In that case we should be able to see the yolk through a transparent shell, right? So let's take a glass and put several egg whites in it together with a yolk. What do we see? We see that the yolk floats. It ought therefore to float inside the eggshell as well.

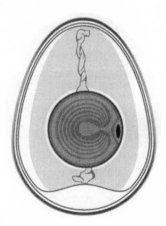

An egg cannot be reduced to a shell, white, and yolk. On the one hand, the shell is a marvelous structure with a scaffolding of organic tissue and striking patterns of mineral crystallization, pierced by pores at several layers. The white is not homogeneous, and while it is mainly composed of proteins and water, here again one finds wonderful evidence of molecular organization. The yolk is made of concentric layers that are successively deposited during the day and the night, in each of which "granules" are dispersed in a plasma. Note also the various membranes, the cords attached to the yoke known as chalazas, and the air pocket at the bottom.

Understanding Phenomena

In addition to testing culinary dictums, molecular gastronomy seeks to model certain transformations. This means we have to understand the phenomena that take place during these transformations—or to put it more precisely, understand the chemical and physical mechanisms underlying the transformations, which in this case are culinary. When an egg cooks, for example, we see that the albumen, which initially was liquid, greenish yellow, and transparent, becomes almost solid, white, and opaque. Why does this transformation occur? To answer this question, we need to be able to explain the accompanying phenomena, or mechanisms. This is the heart of the discipline.

Modeling also frequently involves isolating a fundamental instruction—what I call a "culinary definition"—contained in a recipe and analyzing its implications, which are not always explicitly stated. Consider, for example, the following recipe for stewed pears:

> *Take a dozen pears of medium size, peel them, and put them one by one into cold water. Next, over low heat, melt in a saucepan 125 grams [not quite ¼ lb] of sugar in cube form with a little water. When the sugar is melted, add the pears, sprinkling them with lemon juice if you wish the pears to remain white; if you prefer them to have a reddish color it is not necessary to add the lemon juice. It is indispensable to cook them in a copper-bottomed pan.*

In this case the culinary definition, or relevant instruction, is simply this: Take some pears and cook them in water and sugar. The rest is nothing more than culinary dictums. During the cooking itself, however, one observes phenom-

But, you may object, there are membranes that hold the yolk in the center of the egg. It is true that the egg contains what are called chalazas (the white cords that you sometimes see in the albumen). But do they really serve to center the yolk? If we perform another experiment, removing the top part of the shell of a raw egg with a knife, we will see that the yolk generally floats in the white, just as

ena that the recipe does not describe: The pear pieces soften and the plant tissue becomes transparent. These are the phenomena that we need to try to model.

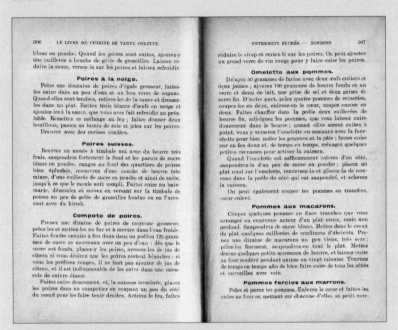

Food recipes combine two forms of intelligence: technical (the basic procedure and the reasons for it) and artistic (why two ingredients with different flavors should be mixed together). Alas, it is usually the case that recipes condemn the cook to being nothing more than an executor. How can this be changed?

it did in the glass we considered a moment ago, in spite of chalazas. This makes sense, of course, since we all know that the yolk contains fats and that these are less dense than the water that makes up the most of the white.

Well, then, how does one center an egg yolk? When you boil an egg, it usually lies on its side in the pan, which means that the yolk floats to the upper part of the

Here an egg cooked to a temperature of 65°C (149°F) sits enthroned upon an asparagus cream sauce.

shell. How do you center it? The solution is simple: Using a wooden spoon, roll the egg around so that no one part of it is always facing upward; when the white coagulates, the yolk will therefore be perfectly centered.

To Be a Cook or a Chemist?

THE PROFESSION OF COOKING

Some people's childhoods seem to determine their choice of profession. Many cooks say they "fell into the pot" when they were young, and whether they loved cooking or whether they felt obliged to follow in their parents' footsteps, they embarked upon a program of apprenticeship or enrolled in cooking school.

There is much to be said in favor of a cooking career. For those who are not mathematically inclined, it requires nothing more than elementary calculations. More importantly, it mobilizes all the senses: smell and taste, quite obviously, but also hearing (cooks often are able to tell from the sound made by a dish simmering on the stove whether it is cooked or not), sight (colors are fundamental in cooking, as classic preparations attest—think of the green of spinach or of the beans that we will prepare later), and touch.

And how can one not like a profession that permits artistic expression? A chef directs the various ingredients at his disposal, much as a film or stage director organizes and shapes the performances of his cast in order to produce a personal statement. A chef also must concern himself with philosophical matters, particularly the question of the "good," which in the case of cooking is none other than what is beautiful to eat. Nor is the chef simply a machine for delivering calories, proteins, lipids, and carbohydrates. He is responsible above all for giving love, for creating happiness. Isn't this a marvelous calling? It is also a complete profession; for the hand of the manual worker will make nothing of value if it is not guided by the head.

THE PROFESSION OF CHEMISTRY

Others received a chemistry set at the age of six, as I did, and were fascinated by the transformations they discovered, especially the one where limewater turns cloudy when you blow air through a straw into it. To perform this experiment, it is necessary first to make limewater. You start by heating calcium carbonate in a very hot oven. This produces a white solid that reacts when you pour water on top of it (whereas the carbonate doesn't react and doesn't even dissolve). What has happened is that the heating has transformed the calcium carbonate into calcium oxide (quicklime), which reacts with the water to form calcium hydroxide (slaked lime). To obtain limewater, you dissolve the slaked lime in water, thus forming a calcium hydroxide solution. Add a spoonful of slaked lime to a glass filled with water, stir, and then filter it the same way you would coffee. When you blow air through a straw into this clear solution, the expelled carbon dioxide reacts with the calcium hydroxide to create calcium carbonate, which then precipitates, producing a cloudy suspension (you can't see the calcium carbonate particles that are dispersed in the water, but they are what makes the water appear cloudy). All these phenomena can be described mathematically in the form of equations—which is itself a marvelous and mysterious thing.

Children find such experiments fascinating, because they show that when we understand matter we are capable of transforming it. To extract metal from ore, to form gas by passing an electric current through water (salted in order to conduct electricity), to distill essential oils from aromatic plants, to create brilliant golden spangles from the precipitation of lead sulfide in water as it cools, to give lively colors to flames with the aid of well-chosen salts—what a delight!

Science museums have done a great deal to introduce science to children. In Paris, particularly at the Palais de la Découverte, demonstrations involving liquid nitrogen have captivated audiences of all ages for decades. At the very low temperature associated with this fluid (-196°C [-321°F]), a leaf from a tree becomes brittle, whereas a very supple plastic pipe becomes hard. If this hardened plastic is tapped with a hammer it breaks into pieces, but if it is allowed to warm up it regains its flexibility. And then there is the dramatic demonstration in which a sort of large matchstick, still glowing red, is plunged into liquid oxygen and catches fire again immediately, illuminating the fog that hangs over the fluid. Children are not alone in finding such spectacles enthralling.

Is chemistry more of a manual trade than cooking? No calculation based on a poorly designed and poorly conducted experiment can be correct. Nor can any dish that is poorly conceived be good. So we should recognize that the so-called manual trades are not inferior to the so-called intellectual professions. A profession (such as chemistry) in which the hand does not play an active role is sadly incomplete, and a trade (such as cooking) from which the head is absent is worthless.

FOR THE UNDECIDED

Finally, there are those who have never been exposed to the decisive experience that determines a vocation. For them, the British scientist Francis Crick, cowinner of the 1962 Nobel Prize in Medicine, proposed a simple test. Crick had been trained as a physicist, but one day, on coming out of a pub where he had been discussing science with some friends, he realized that he had spoken to them passionately about biology rather than physics. Crick therefore decided to change careers—and it is thus, he says, that he was led to discover the double-helix structure of DNA with his friend James Watson. If you're undecided about what to do, try to discover your true passion by listening to yourself in conversation with your friends.

LIGHTING A FIRE

Obviously it is not always easy to discover what most interests children, but activities they find fascinating cannot fail to reveal certain tastes and inclinations. This is one of the reasons why, in 2001, cooking workshops were introduced in primary schools in France.

In a series of laboratory sessions, children not only explore cooking (which naturally attracts them), but become acquainted with a number of related activities, including chemistry, physics, biology, history, geography, and art. Since gastronomy involves all these things, it makes sense to use food to interest children in something other than eating, without their realizing it.

The experiments done in these workshops are described, and supplemented with teaching suggestions so that any adult can conduct them, on the Web site of the Centre Régional de Documentation Pédagogique. The experiments are designed to help children explore their culinary heritage (they make bread, cheese, and wine), learn about traditional practices and techniques, and inquire into the meaning of culinary activity. During the first workshop, for example, the children make one cubic meter of whipped egg white out of a single egg white and learn why the whipped egg white is both white (since the egg white before it is beaten is yellowish green) and firm (since the raw egg white is liquid). What will these kids do when they grow up?

Hard-boiled egg with mayonnaise is a "modern" dish since the mayonnaise sauce itself did not appear until the seventeenth century.

And Now for the Mayonnaise

Classically, a hard-boiled egg is accompanied by a mayonnaise sauce. But why do we serve a sauce with dishes? In this case the reason probably has to do with the fact that the hard-boiled egg is a little dry, and therefore goes down more easily with a lubricating liquid that helps the saliva to do its job, allowing us to swallow the egg without pieces of it getting stuck in our throat. As a general principle, which will be useful in the task of construction that we have set for ourselves here, it makes sense to provide a liquid sauce for dry dishes, and vice versa.

Contrary to what one might suppose, mayonnaise did not always exist. It is not mentioned, for example, in *Le Viandier*, published in the fourteenth century by Guillaume Tirel (better known as Taillevent), and its current name first appeared only about a century ago. The classic recipe does not call for mustard, by the way, which was reserved for remoulade sauces—a practice that is worth reviving in view of how different the flavors of mayonnaise and remoulade are. Let's aim for precision rather than confusion.

Many things have been claimed about mayonnaise: that it thickens only if the ingredients are all the same temperature; that a woman's period can cause the sauce to break; that it fails when the moon is full; that cold is its greatest enemy; that heat causes it to break; that adding boiling vinegar will hold it together, and so on and so on. Why has mayonnaise inspired so many culinary dictums?

It seems plausible to suppose that recipes for fragile sauces—those that are apt to break—are likeliest to give rise to such rules because cooks, when they encounter failure, cannot help but wonder why. Long ago, unaided by the knowledge provided by modern science, they were reduced to guesswork (the influence of the moon, women's menstrual cycles, temperature, and so on), and these hypotheses became hardened into dictums and accumulated over time.

Today, equipped with the knowledge that children are now being taught in their very first days in school, we can begin to explore this ancient cultural fund. Let me repeat: These explorations are within the reach of schoolchildren, with the help of elementary equipment.

WHAT IS A MAYONNAISE SAUCE?

A mayonnaise, according to Carême, is made by mixing together an egg yolk, vinegar, salt, and pepper, and then adding oil to this mixture, drop by drop, constantly whisking it into the sauce. To understand why combining vinegar, egg yolk, and oil—all of them liquids—produces a fairly solid result (at least when all goes well), we must examine the physical chemistry of these ingredients.

An egg yolk is roughly one-half water, the rest being made up by proteins and fatty molecules, including phospholipids, which are analogous to the detergents we use to clean dishes and to the molecules that form the membranes of all living cells.

Vinegar may be as much as 95 percent water, in which acetic acid is dissolved; additionally, it contains a series of molecules that contribute to the taste and odor of the liquid—or, as we should say instead, its flavor, to bring together all the gustatory sensations under one term.

Oil is oil. It is composed of molecules that do not dissolve in water (these are not fatty acids, by the way, as advertising would have us believe, but triglycerides).

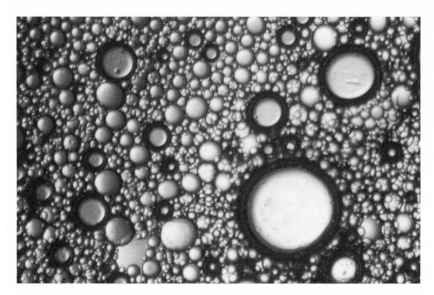

Mayonnaise sauce seen under an optical microscope. When the mayonnaise is made by hand with a whisk, the oil droplets, dispersed in the water of the egg yolk and vinegar, have a diameter of roughly 0.01 millimeter (about four thousandths of an inch).

When one adds a drop of oil to a mixture of egg yolk and vinegar, the oil remains unmixed. Nonetheless, when one beats the mixture with a whisk, the drop of oil is divided into droplets that are coated by the proteins in the egg.

Proteins are like pearl necklaces folded in on themselves, with an inner part that has no affinity for oil and an outer part that spontaneously comes into contact with the water (for fascinating reasons that, unfortunately, would take us too far from our subject). When the oil, yolk, and vinegar are whisked together, the proteins are unfolded and their inner part is exposed in the water. Minimizing their energy, rather like a ball that is released at the top of a hill and rolls down to the bottom, the proteins adhere to the surface of the droplets of oil, coating the droplets and thus stabilizing them in the water. As a result, the droplets find it difficult to fuse with one another. It is sometimes said that such a sauce is stable, but this is not strictly true. In reality, the constituent elements of the sauce separate very slowly (over a period of several weeks).

Why is it necessary to introduce the oil drop by drop, at the beginning of the preparation? Because the objective is to disperse the oil in the water, and not the other way around, which is what would happen if too much oil were added initially; in that case there would be too little water, so whisking the mixture would disperse the water in the oil, rather than the oil in the water, making the emulsion highly unstable. The oil therefore has be poured in drop by drop, to begin with, and thereafter more quickly.

HOW SHOULD ONE WHISK A MAYONNAISE?

The question of how the ingredients of a mayonnaise should be whisked together is an important one, because the answer will determine the final flavor of the sauce. In order for us to be able to smell the result, the odorant molecules must evaporate; they have to escape the sauce before they can reach our nose. These molecules, as it happens, are typically soluble in oil rather than water. Mayonnaises with very small and closely packed oil droplets—a result obtained by using an electric mixer or, following Carême, by crushing them against the side of the bowl with a wooden spoon—have relatively little taste (their sourness is reduced) but a lot of fragrance (assuming one uses fragrant ingredients). Conversely, mayonnaises that are not vigorously beaten, whether with a fork or a whisk, are both less fragrant and less firm, but they are tastier.

WHAT CAUSES A MAYONNAISE TO BREAK?

The most common reason for the failure of a mayonnaise is that too much oil has been added. Since the droplets are dispersed in the water of the yolk and vinegar, there must be enough water to receive the oil droplets. The minimum proportion of water needed is about 5 percent, which, for one egg yolk and a spoonful of vinegar, yields a large bowl of mayonnaise. At this stage, the droplets are so closely packed together that they can no longer move individually; and if none of them can move, the sauce cannot flow. In other words, it is thick. Nonetheless, it is possible to considerably increase this quantity by adding more vinegar. A simple calculation shows that one egg yolk contains enough proteins and phospholipids to make several dozen quarts of mayonnaise.

Note that temperature doesn't matter in making a mayonnaise, any more than the phases of the moon or menstrual cycles. Few kitchens today ever get cold enough for oil to congeal; and simple tests show that mayonnaise can be made with oil as hot as 50°C (122°F). There is no more reason, thus, to insist on the ingredients being the same temperature than to bar women from the kitchen on certain days of the month.

What about adding boiling vinegar to stabilize a mayonnaise? Let's compare a mayonnaise to which water has been added with one to which water (or vinegar or lemon juice) has not been added, whether or not these liquids are boiling. First of all, we note that the addition of a liquid whitens the sauce. The added liquid separates the oil droplets, which had been packed together. Whereas before the light traversed the compressed droplets, the yolk component having been absorbed, now it is reflected by the walls of the separated droplets. The white appearance of the sauce is due to this reflection, not to the absorption of the yolk component. We can see also that the sauce, which before was very firm, has now become a bit looser. A mayonnaise that had been on the verge of breaking because of the very small amount of water present is therefore now more stable. Moreover, it turns out that mayonnaises to which hot or cold water (or vinegar or lemon juice) have been added are equally stable.

A Science for Cooking: Molecular Gastronomy

Thinking about even something as simple as a hard-boiled egg shows that the urge to understand, combined with rational analysis, makes it possible to transform traditional culinary practice. Clearly technology has an important role to play in this. But the role of purely scientific research is even more crucial, for scientific research produces knowledge—the source of technological innovation and the basis for educating future generations.

Molecular gastronomy may seem a pompous name. Nonetheless it is well chosen. Gastronomy is not a cuisine for the rich, as is often supposed; it is what might be called a reasoned discourse about food. This discourse may be historical (the study of how ways of eating have changed over time), geographical (the study of regional variations of dishes), or juridical (the study of laws concerning the labeling and marketing of food products), or indeed literary, or artistic, or scientific. When gastronomy is scientific, it may be chemical, physical, or biological. To designate the specifically physicochemical exploration of culinary transformations, it makes sense to characterize such research as molecular.

THE PROGRAM OF MOLECULAR GASTRONOMY

Cooking, at least in its technical aspect, is studied by molecular gastronomy in two ways, because recipes, whether transmitted orally or in writing, do two things: They give definitions of various dishes (a pot-au-feu, for example, is made by heating meat in water) and culinary dictums about the best way to cook these dishes. It is up to molecular gastronomy to clarify the definitions and test the dictums.

From time to time, obviously, these studies will lead us to refute the great authors of the past (among them Marin, Carême, Urbain Dubois, Édouard Nignon, and Jules Gouffé), especially when these chefs venture to explain the reasons for certain culinary phenomena. Is molecular gastronomy therefore iconoclastic? Yes, in a way. Facts are facts, and errors, even ones committed by the most gifted cooks, are errors. Should we therefore cease to admire the foremost chefs and scientists of the past? Certainly not. The illustrious French chemist Antoine-Laurent de Lavoisier (1743—1794), the father of modern chemistry, believed that all acids contain oxygen. We know today, however, that this is not true (hydrochloric acid, for example, contains only hydrogen and chlorine). And yet chemists will not cease to admire Lavoisier, a remarkably lucid, clear, and intelligent thinker, despite the fact that he was occasionally mistaken—because he cleared a path through uncharted territory for others to follow. Lavoisier pushed the boundaries of knowledge further than any other chemist of his time.

I digress. Simply because molecular gastronomy rejects certain accepted ideas, because it exposes the errors of chefs of the recent or distant past, does not mean that it is an exercise in deconstruction. To the contrary, it is an exercise in renovation—the like of which is long overdue.

Molecular gastronomy makes it possible to place cooking on more secure foundations by recognizing it at last as the chemical art that it quite obviously is. The knowledge it yields has both technological and pedagogical applications, as I say. It is curious, for example, that an egg white should still be said to be made of "albumin," when the chemical understanding of albumin was revised almost a century ago. Lavoisier, in his *Traité élémentaire de chimie* (1789), insisted that science cannot progress if language is not improved. The same thing is true in cooking, where mastery of transformations will be difficult to achieve on the basis of false ideas and of imprecise and misleading words.

Does molecular gastronomy threaten to kill off classic cooking? Ask yourself whether synthesizers are likely to destroy the music of Bach and Mozart. In culinary art, as in musical art, there are technical and artistic components. Knowledge enables artists to refine their technique and helps their art to grow. Nonetheless, it is the artist who decides to work in a different way from his predecessors, not science; it is the artist who bears responsibility for his art, and not science. As for Bach and Mozart, it is up to us to decide whether we wish to keep listening to them or not.

We mustn't forget either that the notion of a golden age is pure fantasy. On the one hand, we are the first generation in the history of humanity not to have known famine—and yet it is only in the West that famine is a memory. On the other hand, whereas nineteenth-century cookbooks very frequently mentioned cases of fraud (watered-down milk, coffee doctored with liquid manure to enhance its color, flour mixed with plaster, and so on), never have we been so well protected by government regulation and never have food products been so safe. Wines are better today than they were even twenty years ago, when the addition of sulfur dioxide still caused dreadful headaches; meats are all the better for being kept in the refrigerator, rather than being left out where flies can get at them. We must listen skeptically to the claims of those who go on about how good life was in the old days—even if, like everyone else, we harbor a nostalgia for a past that never really existed.

Simple Consommé

A consommé is a bouillon that is served, hot or cold, at the beginning of a meal. Simple consommé—a clarified broth made by cooking meat or fish in water—is a staple of home cooking, sometimes supplemented by vermicelli, tapioca, a julienne, croutons, or something else. Double consommé is a more elaborate preparation whose flavor has been enriched by reduction and further clarification.

The chef Jules Gouffé (1807–77), whose cookbooks are notable for their remarkable precision, held that "bouillon is the soul of home cooking."

Yes, it is the soul of home cooking, because it is a liquid that is served every time a pot-au-feu is made; but it is also the foundation of more sophisticated dishes, because it forms the basis for sauces, stocks, demi-glaces, jellies, consommés, and so on.

Many recipes can be made from a bouillon. No dish in a classic cuisine worthy of the name would have been moistened with anything other than a good bouillon: fowl, veal, game, fish, vegetables. In fact, bouillon is nothing more than flavored water—but home cooks sometimes forget that cooking is a matter of giving flavor to foods, not of preserving their "natural" flavor.

In constructing the dishes that constitute our meal, this is an important point, and one that has its counterpart in all the other arts. Musicians? They don't reproduce the sounds of nature; they organize individual notes to evoke the sounds of nature more movingly than nature itself. Painters? They do not reproduce nature; they compose, construct their canvas, creating a foreground and a background, embellishing the image. Photographers? Open any magazine and look at any photograph, and you will see that the aim is not to reproduce what we ordinarily see, but to produce a view of what the photographer wishes us to see. The same is true in literature, in sculpture, in cinema. Generally speaking, amateur photography, amateur painting, amateur music are not very good. Why should amateur cooking be any different?

Professional cooks understand that to give something the flavor they think it ought to have often requires a lot of work, preparation, and seasoning. Pierre Gagnaire, for example, speaks of setting the stage for his ingredients, not unlike a film or theater director.

A Site for Sharing New Ideas

My collaboration with Pierre Gagnaire, the results of which are available on his Web site (www.pierre-gagnaire.com, "Art et Science"), began in 2000. In November of the preceding year, Guy Ourisson, then president of the Académie des Sciences in Paris, invited me to give a lecture there, accompanied by a meal based on the principles of molecular gastronomy. At the same moment, as it happened, Pierre Gagnaire's wife suggested he invite me to help him prepare a special menu for the celebration of the upcoming millennium.

By then Pierre had left Saint-Étienne, where in two different restaurants he had established a reputation for culinary genius. He opened a restaurant in Paris, near the Champs-Élysées on the rue Balzac, that was immediately awarded three stars by the Michelin Guide. There he continued to develop his artistic talents with unremitting originality.

And so it came to pass that the dinner at the Académie des Sciences was planned and prepared by the two of us. A series of working sessions led to a wonderful meal, as well as a new menu featured in the winter of 2000 at Pierre's restaurant—a splendid example of science and art working hand in hand.

We met a few times after that—social encounters, mainly, but nonetheless full of talk about the technique and art of cooking—and finally decided to set up an Internet site where each month we would post an "invention" based on research in molecular gastronomy together with artistic interpretations. I would write a description of the new idea, which Pierre would illustrate with four recipes.

In this way we hoped to follow in the footsteps of the French chemist Michel-Eugène Chevreul, whose work on color a little more than a century ago led to the creation of neo-impressionism, the leading representatives of which included the painters Paul Signac, Camille Pissarro, and Robert Delaunay. Today, the enterprise of placing molecular gastronomy in the service of culinary art is eight years old, and each month fresh applications of research continue to be added to the site's archives. Who says French culinary creativity is dead?

How to Make a Bouillon

Making a stock? It's so simple that it hardly seems worth explaining. One puts meat in water and heats it. Ah, but what sort of meat? From what part of the cow, if it is a beef bouillon? Fresh meat or meat that has been aged? And how much meat for how much water? What kind of water? Salted? Heated in what sort of pot? Earthenware? Cast iron? Stainless steel? Covered or uncovered? Cooked at what temperature? For how long?

HOT OR COLD WATER?

The number of culinary dictums I have collected relating to bouillon is far greater than the number relating to mayonnaise, for example. One finds all sorts of instructions. Some recipes say that the cover of the pot containing the bouillon should be left ajar by the width of two fingers. The majority say that the meat must be immersed in cold water, never in boiling water; but there are nonetheless those who claim the opposite. The cooking time varies, depending on the author, between one and twenty hours—which doesn't give much guidance to the home cook. Some say the cooking temperature must be low; others say that the liquid should never stop boiling, but that, if it does, it should be brought back to a boil gently. Every imaginable recommendation can be found if you look hard enough; and science only confuses matters further, unfortunately, by introducing mistaken notions.

Whereas L.S.R., an anonymous author of the seventeenth century (he is said to have been called "Le Sieur Robert," but this does not help in identifying him) insisted upon placing the meat in boiling water in order to make a good bouillon, many nineteenth-century authors recommended instead beginning with cold water. Surprisingly, the German chemist Justus von Liebig (1803–73), despite his great eminence (he was knighted for his remarkable feats of chemical analysis and for the institute he founded at Giessen, whose students went on to make his influence felt throughout the world), endorsed the majority opinion without offering any justification for it. "If the meat is put into boiling water," Liebig summarily stated, "the albumin coagulates on the surface and prevents the juices from escaping and making a good bouillon."

Why should we doubt Liebig? Because the mission of science is to doubt, and also because the counterindication given here, subsequently repeated by French authors (as a young man Liebig had learned French in the kitchens of Louis I, Grand Duke of Hesse-Darmstadt, before going to study chemistry in France), is, well, doubtful. In fact, meat is not chiefly composed of albumin; or, more to the point, albumin is not what is at issue in this case.

To see why this is so it is necessary to appreciate the fact that meat is a muscle tissue, composed of muscle fibers, which is to say living, elongated cells with contractive properties. When the brain sends the signal, proteins within the fibers (mainly actin and myosin) slip past one another, shortening the muscle and causing it to contract. Additionally, the fibers are sheathed with a tough reinforcing tissue formed from a highly fibrous protein, collagen. The muscle tissue itself also contains fatty deposits and a circulatory system. The blood in this system conveys a number of different proteins, including small globular plasma proteins called serum albumin. There is therefore some albumin in meat, but very little.

Why, then, did Liebig mention only albumin? Because in his time this was a general term, used to refer to the entire class of proteins. The first proteins to be identified were found in eggs. At the turn of the eighteenth century, the French chemist Étienne-François Geoffroy (1672–1731) gave the name *albumin* to the proteins of egg white (or, as we now say, albumen). At the time, albumin was quite a mysterious substance, detectable only by its power to induce coagulation, its tendency during putrefaction to release nitrogenous gases (nitrogen had just been discovered), and its ability to impart a violet color to syrup. Geoffroy soon recognized that certain plants, such as peas, also contained products of the same type, which were called "plant albumin." This was a revolutionary discovery, because now it was understood that the same molecules occur in both the plant and animal realms. Later on, improved methods of chemical analysis made it clear that egg white was composed of many different compounds, and that collagenous tissue, although it contained proteins, was not coagulable, unlike tissue containing actin and myosin. In other words, egg white was not made simply of water and albumin. Gradually the term *proteins* gained currency, with *albumin* being restricted to a class of small water-soluble proteins. Thus eggs contained ovalbumin, and meat, serum albumin.

A SIMPLE TEST

Scientific claims formulated in terms that were current more than two centuries ago are very likely to be inaccurate. And this is why Liebig's account needs to be tested. The relevant experiment is not difficult to perform. Take two pots and fill them with the same quantity of water; bring the water in the first pot to a boil, without heating the water in the second; then put the same amount of meat in each, turning on the heat under the second pot while keeping the first one at a boil, and weigh the pieces of meat in the two pots every minute. What you will find is the opposite of what Liebig had predicted, namely, that after an hour of cooking the meat that was initially placed in boiling water has lost more mass than the meat that was placed in cold water. Eventually, the two pieces of meat come to weigh the same amount, to within a gram of each other, and this mass does not vary no matter how much longer the meat is cooked. The inescapable conclusion, then, is that the meat loses mass during cooking mainly because the collagenous tissue contracts when it is heated. And because it is heated more rapidly in boiling water, it contracts to a greater extent, initially forcing a larger quantity of water into the bouillon. Ultimately, however, it doesn't matter whether one starts out with hot water or cold water.

The flavor of a bouillon comes chiefly from the slow disintegration of the collagenous tissue. This produces gelatin, which then dissolves in the bouillon. Gradually, the gelatin is degraded by the heat into very flavorful amino acids. Odorant molecules are released as well. These molecules absolutely must *not* be allowed to disperse into the air and disappear—which is why one must make sure that bouillons are not brought to a rolling boil. All chemists are familiar with the technique of extracting water-insoluble molecules, such as those of essential oils, by separating them in gaseous form from water vapor. This separation process has to be avoided at all costs in making bouillon. Hence the importance of keeping a lid on the pot and keeping the temperature low, which will allow the collagen to dissolve and break down into its constituent amino acids, without losing any of the bouillon's wonderful smell. The cooking time? Quite a few hours are needed to extract the maximum amount of gelatin from the meat and to give this gelatin and other proteins time to hydrolize, which is to say break down into amino acids.

A Technical Question

Clarification, which turns ordinary bouillon into a simple (or double) consommé, shows how anachronistic our practices are and how useful applying the principles of molecular gastronomy can be.

The point of clarifying a bouillon in the first place is to make it clear. The turbulent action of boiling the meat frequently extracts particles that make the bouillon cloudy. Clarification produces a limpid, amber liquid, wonderful to behold and delicious to consume. Classically, one begins by allowing the bouillon to rest so that the fat, suspended in the liquid in the form of droplets, has time to rise to the surface, where it congeals in a layer that is easily skimmed off with a spoon (in the old days this fat was kept in a "grease pot" and reserved for further cooking). Elementary physics shows that the speed with which these fatty droplets rise to the surface depends on their size; the smaller the droplets, the more slowly they rise, although they are the ones that congeal first. This means that the bouillon ought not to be cooked too quickly, if you wish to be able to remove a well-formed layer of fat from the surface of the liquid.

THE STAGES OF CLARIFICATION

Once the bouillon has been degreased, it is clarified in several steps. You begin by decanting, in order to isolate the clear liquid in the upper part of the pot, tilting it so that the particles that cloud the bouillon remain undisturbed at the bottom. This first batch is then filtered. But since filtration in its classic form makes use of a chinois (or "China cap") lined with cheesecloth, some particles remain suspended in the filtered liquid. It is therefore necessary to perform a second clarification, by adding egg whites to the cold bouillon, then whisking the mixture and gently heating it until the albumen coagulates around the suspended particles. The bouillon is then strained through a chinois, or other strainer, lined with a piece of clean cheesecloth that has been folded in four to filter the liquid more finely.

This procedure has the disadvantage, however, not only of wasting egg whites but also of weakening the flavor—which is why conscientious chefs enrich the bouillon by adding, along with the egg white, chopped meat and vegetables.

What a complicated business! Wouldn't it make more sense simply to extract the flavor molecules of the initial ingredients and then properly filter the reserved liquid? Chemistry laboratories are full of devices that rapidly clarify fluids—fritted glass filters, for example, made out of compressed glass particles, with perfectly calibrated pores that are sufficiently small to block the passage of all the particles that cloud bouillons. And if the filtering process seems too slow, there are devices that create a vacuum, sucking the liquid through the filter and producing a clear bouillon in a few seconds. Isn't it time that we made use of such devices, rather than continuing to rely on a chinois and cheesecloth?

This is a good example of a technological application of molecular gastronomy. It is not a question of science, but a way of rationalizing culinary procedures with a view to facilitating artistic endeavor. At first, during the formative period of molecular gastronomy, few people realized that this business of modernizing traditional practices was not scientific in nature, but technological. Things are different now, and it is important to point out that the cooking profession has adopted many of the modern techniques that have been proposed—though not yet the wonderful method of filtration I have just described.

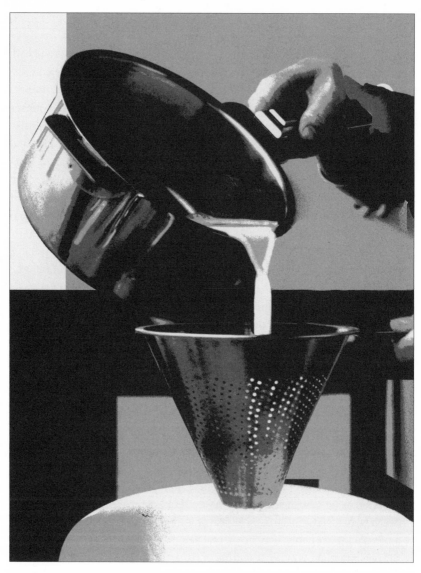

The chinois is an inadequate culinary instrument, because its holes are too big to trap the very small particles that cause bouillons to turn cloudy. What if one were to redesign the equipment used for clarification?

Cooking Has Really Changed!

Until fairly recently, mayonnaise was made by rote, following a simple set of instructions: Take one egg yolk; add vinegar, salt, and pepper, then whisk in oil. Modern methods of physiochemical analysis brought a new way of thinking about mayonnaise. They made it plain for everyone to see that the sauce is in fact a dispersion of microscopic droplets of oil in water, supplied by the egg yolk and vinegar, and settled once and for all the question of how to save a mayonnaise that had broken.

The implications were huge, because by providing visual evidence of oil droplets packed together in the sauce, electron and light microscopy forever deprived culinary empiricism of its innocence. No one who saw these droplets with his own eyes could ever forget them. This is a good example of the truly revolutionary impact of science, which, by producing new knowledge, shakes up our view of the world and alters our behavior.

Yet there were some who concluded that, since mayonnaise is an emulsion, all you have to do is disperse oil droplets in water. But how exactly are they to be dispersed? Carême used a wooden spoon, pressing it against the side of the bowl in which the sauce was made and in this way crushing and dividing the oil drops. What a lot of work—making mayonnaise took an energetic quarter of an hour. Then the whisk, in the old days made of wicker and today of stainless steel, replaced the wooden spoon. Next came the electric mixer. Was the mixer really better suited to the job? No one seems ever to have posed the question, for the simple addition of an electric motor led in short order to the food processors we know today. And yet if we had stopped to consider that a drop of oil added to the sauce and divided in two by each revolution of one of the wire loops of a whisk produces a large number of very small droplets, we would have realized that the number of loops is the main thing. The more loops there are, the more efficiently the droplets are divided. Moreover, two whisks will do the job twice as quickly as one.

Better still than the whisk or the food processor is the sonicator, a device used by physics and chemistry labs everywhere to make emulsions. It emits ultrasonic waves, or ultrasound, like the tanks that are now used to clean eyeglasses. You put the ingredients in the sonicator, push a button—and bingo, it's done! But do we really have to give in to this tempting new device? (Interestingly, the etymology of the word *engine* is connected with the notion of the devil.) Can't we go on using the wooden spoon? The very question is telling, since to pose it is to answer it. No sooner will the world of cooking have learned of the existence of sonicators than someone will go out and buy one, in the hope of saving some time, and put it to good use. As a result, mayonnaise will be transformed, because its physicochemical structure will have changed. The size of the oil droplets varies depending on whether one uses a wooden spoon, a whisk, or a sonicator. And we know today, at least in this particular case, that the flavor of the sauce will change as well. Mayonnaises with large oil droplets have a more pronounced fragrance of egg yolk and vinegar than ones with very small droplets, which bring out the characteristics of the oil to a greater degree.

Let's come back to the question of how best to thicken a mayonnaise. From the foregoing it is plain that science, technology, and technique all need to be taken into account. Science searches for the mechanisms of phenomena in order to produce knowledge, and nothing but knowledge. Technology uses this knowledge to improve technique—and technique is what we employ to make things.

In cooking, technology has lagged far behind, because without a science of culinary transformations there could be no new knowledge, and therefore nothing to apply. Technique changed only slowly, as a function of experience. Nor have these distinctions always been clearly understood. At the time of molecular gastronomy's founding, two of its objectives were to invent new dishes and to introduce new methods, utensils, and ingredients. Both of those aims therefore involved technology. This may be the reason for the misapprehension one still finds both in kitchens and in the press, where molecular gastronomy has been confused with "molecular cuisine" or "molecular cooking"—a culinary trend that is the result of applying molecular gastronomy, in particular through the introduction of new tools, ingredients, or methods.

Unlike discovery, which comes under the head of science, invention is the prerogative of technology. Invention in cooking is simple, so long as the cook has the right intellectual tools. Thus, for example, a mathematical formalization worked out in 2002 to describe the physicochemical structure of foods has made it possible to conceive of an infinite number of new dishes. I invented a device called a "pianocktail" (the term was coined by the French writer Boris Vian) that produces examples of complex dispersed systems on the basis of the formulas generated by the mathematical description. Volker Hessel and his team at the Institut für Mikrotechnik in Mainz, Germany, constructed a prototype of this machine. Controlled by a computer and connected to a network of pumps and microreactors, the pianocktail makes cocktails and various other preparations (meringues, sauces, suspensions, gels, and so on) when a formula is typed into the computer. A system of microreactors within the device makes it possible in principle to produce more than 500 billion different complex dispersed systems—not counting the possibilities for manipulating the flavor of these systems.

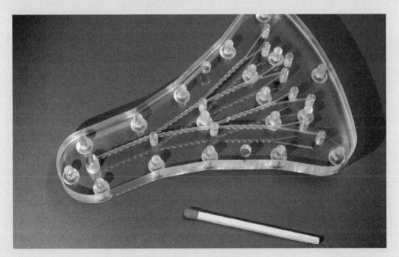

This miniature device, composed of "microreactors" that generate instructions for combining liquids, powders, and gasses in various proportions, makes it possible to produce hundreds of billions of new dishes.

Whereas cooking in the past was condemned by its reliance on repetition and the mechanical execution of familiar recipes to use traditional utensils, molecular gastronomy opens up the possibility of specifying a particular culinary function before deciding on the precise tool needed to carry it out. Say you want to make a foam. In that case you need to inject air bubbles into a liquid. You can do this with a siphon, the device used today in the most fashionable kitchens, but you could just as well use a compressor or a bicycle pump. Or perhaps you want to make an emulsion, or grind something up, or heat something, or filter something. Laboratory equipment catalogs run to thousands of pages, and you can be sure of finding just the right tool for the job.

Thirty years ago, such ideas would have been dismissed out of hand. But times change, and resistance to modernization has begun to diminish. It took almost twenty years after the advent of molecular gastronomy for liquid nitrogen to come into general use in producing ice cream and sherbets. At the opening session of a recent culinary equipment exposition in Strasbourg, five European chefs demonstrated these and many other uses for liquid nitrogen, as well as ways for using ingredients once considered impossibly exotic, if not actually diabolical, but which are now found in a growing number of cuisines.

Prior to the mad cow disease crisis of the 1990s, for example, the preferred culinary gelling agent was gelatin extracted from calves' feet, either in leaves or powdered form. Gelling agents extracted from algae—agar, carrageenans, alginates—were rejected; in Europe, their code names (E401, E406, E440) gave rise to irrational fears of toxicity. Then came the mad cow scare and gelatin was suspected of contamination (wrongly, as it turned out); but, with hardly any qualms, chefs substituted the very additives they had previously ignored—so widely, in fact, that one of gelatin's most attractive properties (it melts at 36°C [97°F], which is to say in the mouth), is now almost forgotten.

Must we change the way we cook? I believe we must. We can no longer afford the luxury of living in a world in which gas and electric burners waste up to 80 percent of the energy consumed, when induction stoves, microwave ovens, and other energy-efficient systems are now available. What is more, our outmoded cooking habits endanger the world in which our children will live.

Surely it is our civic duty to reintroduce the teaching of cooking in school, not in the form of mechanistic protocols that demean the person who executes them, but by relating it to science and art—in a word, to culture.

In cooking, no less than in other fields, we will forge the world of tomorrow today. Science, in its indispensable relation to technology and technique, has a vital role to play.

Constructing and Learning

In the preceding pages it may have been forgotten that our aim here is to construct, not to explore. If we wish to make a consommé, how should we go about it? Studies in nutrition and physiology give useful hints.

Research on satiety, for example, shows that the sensation of being full occurs when at least two conditions have been satisfied: The flavor receptors have been sufficiently stimulated, and a sufficient amount of time has elapsed to allow their signals to be processed by the central nervous system. Chefs who care about their guests' waistlines and health must therefore seek to stimulate all our taste receptors. What are they?

STIMULATING ALL THE SENSES

The taste receptors of the papillae are sensitive to many molecules. Our mouths, and particularly our tongues, detect not only the molecules that give the familiar tastes of sweet, sour, bitter, and salty—which, owing to our peculiar stubbornness, are still known as the "fundamental" or "basic" tastes—but also molecules that yield tastes other than these four. Knowing, as we now do, that the theory of four basic tastes is false, why shouldn't we use ingredients such as licorice, which contains glycyrrhizic acid, a distinct taste that is neither sweet nor sour nor bitter nor salty; or shellfish, which communicate the taste of glutamic acid and alanine; or sugars such as glucose, whose taste is less sweet than sucrose?

Nor must the confection of a good bouillon fail to stimulate the olfactory receptors. This requires skillfully using fragrant ingredients in order to fill our

Often we perceive the outer surface of a dish more clearly than what is inside. A dish composed of a creamy filling (whipped cream or chocolate mousse, for example) between two wafers seems hard because the teeth come into contact first with the wafers. Whence the general idea of changing the surface of dishes, as in the case of this dessert, where the creamy interior is held together by a supple and slightly elastic sheath.

bouillon with as many odorant molecules as possible—something that is easier said than done, because many odorant molecules are insoluble in water, and bouillon is mostly water. Why not try enriching the bouillon with gelatin, which, by binding with the odorant molecules, will help retain them in the liquid, and so lengthen its flavor in the mouth?

We need also to think of the receptors of the trigeminal nerve, which descends from the rear of the brain and radiates throughout the nose and the mouth, where it serves to detect cool and spicy sensations. Spiciness is classically contributed by black pepper (whose pungency is chiefly due to the action of a molecule known as piperine) and various hot peppers (whose main active component is capsaicin); the impression of coolness is due to a variety of molecules whose mechanisms unfortunately are still poorly understood. Nonetheless, cooks have a large body of empirical knowledge that it would be a shame not to put to use, even if it remains mysterious. They know very well that cucumber is perceived as cooling; red bell pepper, too, even when it has been cooked for a long time. They are also familiar with the refreshing properties of herbs such as basil and chervil, and of the bracing smell of thyme and lavender and mint.

Cooks who wish to arouse the senses as fully as possible, mindful that rapid satiety will prevent their guests from gorging themselves, look also to stimulate the visual receptors. Classically, bouillons are colored with the aid of caramel or burned onions for this purpose, but there are many other possibilities. And then there are the tactile receptors, which detect consistency (the crunchiness of croutons in a bouillon, for example), and thermal receptors, which are stimulated by the juxtaposition of hot and cold elements (you might imagine putting cubes of a frozen liquid in a hot bouillon), and so on.

WHY DO WE FIND EATING PLEASURABLE?

In discussing the question of satiety one can hardly avoid mentioning the experiment in which a balloon is inflated inside the belly of a rat. The rat continues to eat because its taste receptors, which communicate with the brain and tell the organism when to stop eating, have not been stimulated. This type of experiment raises fascinating questions. Why do we take pleasure in eating? What is hunger? Is satiety the opposite of hunger?

It is important to note that some ten to twenty minutes elapse between the moment when we first detect taste or odorant molecules and the moment when we stop being hungry. In the case of our consommé, we need to make a point of slowing down our rate of consumption, by whatever means necessary, in order to take up the full twenty minutes our body needs. (Jean-Antheleme Brillat-Savarin [1755-1856] was quite right, as we can see today, to rail against gluttonous eating.) Experience has taught cooks various tricks, such as garnishing a plate with stems of chervil or parsley that need to be chewed. By helping to slow down the rate of intake, they draw the diner's attention to flavor molecules that would otherwise go undetected, and so produce a more agreeable sensation of satiety. It therefore makes sense to serve a consommé, especially one that contains bits of this or that, at the beginning of the meal. Come to think of it, why shouldn't we serve it before the hard-boiled egg with mayonnaise?

Leg of Lamb with Green Beans

Here again I propose a familiar dish—almost a cliché, in fact, at least in the Western world. A leg of lamb with green beans calls to mind Sunday dinner, especially at Easter. Why should we think of serving this dish when we want to construct something new? Because when we eat, we're consuming not only lipids, proteins, and carbohydrates—we're also consuming culture.

Flavor and Culture

Munster, for example, is a cheese that, in spite of its delicate flavor, has a strong smell that enchants people from Alsace and anyone else who has grown up with it. But people who were not exposed to it when they were young find the smell disgusting. Then again, few Westerners can imagine eating steaming-hot monkey brains out of a skull that has just been cut open. Neither dish poses any danger to our health, but they don't please us equally. It is all a question of culture, of being comfortable with what one is already familiar with.

When the military pharmacist and agronomist Antoine-Augustin Parmentier (1737–1813) tried to introduce the potato in France, as a way of offsetting wheat shortages, he encountered severe resistance, and only with the support of the king was he able to persuade people of the potato's virtues. Still today there are people who go hungry, even though there are edible vegetables nearby, because these vegetables are not part of their food culture. The reason that we do not eat

A sprig of chervil, parsley, basil, mint, or thyme placed on top of a dish is a sort of advertisement that alerts diners to expect a specific taste or flavor.

everything that is available to us is because our culture often prevents us from trying unfamiliar things that we might like.

At the same time, we serve our guests dishes they recognize in order to please them, for such dishes indicate our membership in a group. It is not often enough stressed that as human beings we are sociable animals, like our fellow primates, and that this sociability is a direct consequence of the release of molecules in the brain that give us pleasure when we are in a group. We are human, to be sure, but there is nonetheless an animal part of our nature. And so it is a good idea, if we wish to prepare a meal our guests will enjoy, to serve a dish they are familiar with, such as a leg of lamb with green beans.

Culinary Innovation

What counts as an innovation in cooking? A recent INRA (Institut National de la Recherche Agronomique) molecular gastronomy seminar devoted to this question concluded that an innovation is a successful novelty. In business, one encounters the related idea that a novelty becomes an innovation when it becomes marketable. Both these answers are fine as far as they go.

But the question of innovation is better posed when we ask, "Are there rules for composing new dishes that will be generally accepted?" Inevitably this recalls Jean Cocteau's remark, "Since these mysteries are beyond us, let's pretend that we're the ones behind them." It's always easy, once an enterprise has succeeded, to pretend that one knew all along that it would succeed. Ought we not, more modestly, acknowledge the importance of hard work? If one enterprise in ten succeeds, someone who launches one hundred enterprises will end up with ten that succeed, whereas a person who launches only eight will meet with no success at all. Sensible people seldom dare to predict how such things will turn out. Trying is what matters. After all, as the old proverb has it, "Nothing ventured, nothing gained."

Let's come back to the question of rules for composing new dishes that have a chance (rather than a certainty) of being liked. If we don't sit down at the piano, we won't have any chance of creating a melody. This simple observation shows the importance of studying technique: First of all, we have to learn to play the instrument. The INRA molecular gastronomy seminars in Paris, which are open to anyone who wishes to explore cooking, serve their purpose if they help cooks have a better grasp of the technical aspects of their work. Nonetheless, technique counts for nothing without what I call culinary art, which is the intuition that leads to the creation of new dishes that will be widely accepted. An artist such as Pierre Gagnaire, one of the most creative chefs in the world, is not always sure why he comes up with certain combinations, why he treats certain ingredients in one way rather than another. It is almost as if the idea finds him rather than the other way around, because he has been passionately devoted to cooking for decades, immersed in the art that is peculiar to it and constantly searching for novel flavors. If you look at the "Art et Science" section of his Web site, where the results of our collaboration are posted, you will see that the explanation for a certain choice often comes long after the choice has been made.

One might suppose that novelty in cooking is forbidden because, like monkeys, we are primates. Ethologists have demonstrated that monkeys exhibit a behavior known as food neophobia, which is to say a fear of new foods. They do not eat what they do not know. Such a behavior is easily explained in evolutionary terms: A primate species that ate anything and everything would have consumed toxic vegetables and disappeared long ago. As primates ourselves, we have been saved from extinction by this instinct. There is also the phenomenon of aversion, or a rejection of any food that has already triggered a negative reaction from the organism. Here again, we are very much animals. Who among us hasn't gotten sick after eating oysters or some other food (something fried, a chocolate mousse, or whatever) and then stayed away from this food, at least for a time?

But dietary behavior has other origins as well. Our intellectual faculties probably explain why we tend to eat within our own culture. Why does an Alsatian like Munster and an Indonesian like durian (a fruit whose odor Westerners find revolting)? Why does the Breton cherish his crêpes whereas the Provençal adores garlic? Because culture, and in particular childhood recognition of some dishes as part of this culture, teaches them to value such things. Every departure from one's culture is a piece of daring, a deviation—indeed, a perversion.

Considered in this light, culinary innovation seems doomed to failure. Yet Brillat-Savarin rightly said that "the discovery of a new dish does more for human happiness than the discovery of a star," and chefs, as we know, are forever trying to come up with new flavors, combining basil with lemon and veal with tuna and so on. How is the contradiction to be resolved?

Let's keep in mind, first of all, that omnivorousness has certain biological advantages, because animals that limit themselves to a single food will not survive if it can no longer be obtained. Biological evolution has selected for a behavior that counterbalances the neophobic instinct, so that we are interested in searching for new foods, at least up to a point.

There is another countervailing tendency as well, what ethologists call the "beer and tobacco effect." In humans, as in monkeys, membership in a social group encourages an appreciation of foods that would otherwise be physiologically rejected; it is mainly in order to enjoy the pleasures of belonging to a group that human beings, who as infants exhibit a strongly adverse reaction to all bitter tastes, come later in life to like drinking beer and other alcoholic beverages.

Neophobia is further offset by a human capacity to seek and analyze "forms"—not geometrical forms in this case, but specifically culinary forms. One finds much the same thing in music. A child exposed to free jazz has little taste for it at first, but if he gradually works his way from Louis Armstrong to Duke Ellington and then to John Coltrane, it is possible that one day he will enjoy listening to Youssef Lateef.

Analogies are sometimes misleading, of course, but a similar change in attitude can be detected in recent decades with regard to novel food combinations. Rosemary chocolate candies, which only twenty years ago seemed an abomination, are now appreciated by even the most conservative gastronomes, and the pairing of strawberries and Camembert, which seemed to be a daring piece of iconoclasm in 1913 when it was first suggested by the French chef Jules Maincave (who was killed in the First World War), today seems rather tame, at least in certain restaurants. To some extent this is surely due to a gradual process of habituation. But subtlety may be required as well. Cooks say that for a tarragon sauce to be successful, for example, one should have to search for the flavor of tarragon. In other words, too prominent a flavor is apt to ruin the effect. One finds a similar phenomenon today in the world of wine, where "subtle" wines are increasingly admired.

Innovation is therefore indispensable, but only if it takes into account the physiological, biological, and cultural context of cooking. Hence the importance of acclimatizing ourselves to new ideas, of rooting them in what we are already familiar with. It is for this reason, for example, that I suggested calling a chocolate mousse made by whisking chocolate into a fragrant and flavorful water solution "Chantilly chocolate" (see page 100).

To summarize a discussion that has only just begun: A new dish will win general acceptance only if it is good. What is the good? What is the beautiful? The answers will not come from engineers, but from artists. And they will not come in the form of algorithms, but in the form of actual dishes, which will then be produced for mass consumption. To each his own special talent. May the arts and sciences forever be allied with one another.

To Grill or to Braise

Once again we have strayed from the business at hand. The time has come to cook our leg of lamb. How can we make sure the meat will be juicy and tender? Supermarkets offer meats for grilling and others for stewing. In both cases we are cooking the same thing: muscle fibers bound together by collagenous tissue. The proportion of this tissue, which determines the meat's toughness, varies depending on the type of animal or fish. Most fish contain very little collagen; a filet of beef contains a bit more; and a neck of beef much more. Unfortunately, tender cuts of meat are expensive, whereas the tougher cuts are often the ones that have the most flavor.

From the perspective of classical cuisine, the cook is limited to bringing out the best qualities of a piece of meat, trusting his butcher to have chosen the meat well in the first place and then to have aged it so that it is properly tenderized. To choose a piece of meat, pinch it between your fingers. If they sink in easily, the meat is tender and suitable for grilling; if not, it should be braised.

Is it inevitable that most people will go on buying costly cuts for grilling while neglecting cheap braising meats that are more compatible with family budgets? Everyone likes hamburgers, but few people realize that the technique of mincing meat was originally invented to make the most of tough cuts. Destructuring the muscle tissue made the meat easier to eat, in addition to being less expensive. And when you're tired of hamburger, you can always use a knife with a very thin, sharp blade to cut slices of carpaccio.

But perhaps the best way to treat tough meats is to braise them. Classically, this involved browning the meat in a covered baking dish (originally a brazier, nowadays a casserole) in order to kill any pathogenic microorganisms that might be present on the surface (though obviously this isn't what cooks way back when thought they were doing), then cooking the meat very slowly in the brazier with a bit of liquid (bouillon, wine, or brandy, for example)—"ashes above and below," as it used to be said (referring to the embers that were placed on top of the shallow lid of the brazier, which itself was placed over a wood fire). A leg of lamb cooked in this way for seven hours (even as many as eleven hours, one is told) was meltingly tender.

Why? Because the collagenous tissue, the source of the meat's toughness, breaks down slowly, producing gelatin. As we have seen, gelatin breaks down into amino acids, which powerfully enhance the taste (hence the flavor) of the sauce. Several hours of cooking are therefore required (as in making a bouillon), at a very low temperature (as in the case of the 65°C [149°F] egg). This is why cooks in earlier times feared flare-ups when they were braising meats, for a sudden excessive rise in temperature threatened to ruin the result. Today, why shouldn't we take advantage of modern utensils? A heat gun (the kind you use for stripping paint) will give the meat a nice brown color, which is a sign of marvelous flavor; then cook the meat in a covered casserole in an oven set at 70°C (158°F) for several hours, after which it will be remarkably tender.

Is this the future of cooking? Probably so. More and more professional cooks over the past few years have adopted a technique known as low-temperature sous vide cooking, which, as the name suggests, has two aspects. First, low temperature. This is what we have just been talking about—nothing new here, only the heat is better regulated owing to the improved technology of modern ovens. Second, *sous vide* (a French phrase meaning "in a vacuum"). The meat is placed in a plastic bag in which a vacuum is created, allowing the tenderized meat to be cooked in a closed environment without any loss of its sapid and odorant molecules.

The Green Color of Green Beans

While the meat is simmering in its casserole, let's turn our attention to the green beans. Unfortunately, our methods for cooking vegetables are still rudimentary. There is no reason why cooking them in boiling water (classically known as *cuisson à l'anglaise*) should give them flavor—quite the opposite! This method ought to be used only to blanch vegetables. Immersing them in boiling water for just a few moments is enough to inhibit the action of enzymes, particularly those

This bunch of green beans owes its green color to chlorophyll molecules (there are several kinds having different colors) and to carotenoid molecules, which impart a reddish yellow shade.

responsible for fruits and vegetables turning brown (think of apple slices that are exposed to the air for more than a few minutes), known as polyphenol oxidases.

This term designates a class of molecules that we have already discussed: proteins. But polyphenol oxidases do not function as building blocks in plant and animal organisms; instead, they are chemically active proteins that function more as workers. The suffix "-ase" indicates that they are enzymes; the rest tells us that they are enzymes that oxidize polyphenols. Where do these polyphenols come from? They are very abundant in plants. Some are pigments, typically red or blue, found in fruits and flowers; others are tannins; still others are precursors of odorant molecules.

In plant cells, polyphenols are segregated from polyphenol oxidases in separate compartments. Things take a turn for the worse when a fruit falls to the ground or when a cook cuts up a vegetable. In the first case, the fruit is bruised, which means that the cell walls are damaged and the enzymes come into contact with the polyphenols; in the second case, the compartments severed by the knife likewise bring the two types of molecules, polyphenols and enzymes, into contact with one another. In both cases, a chemical reaction takes place that forms highly reactive molecules known as quinones, which combine to form brown compounds similar to the ones that appear in our skin when it is exposed to the sun.

To prevent sliced vegetables from getting a tan, so to speak, it is necessary to inhibit the polyphenol oxidases. Cooks rightly concluded from experience that lemon juice, which contains vitamin C (chemically known as ascorbic acid), is an effective inhibitory agent. But in that case why shouldn't we buy some vitamin C tablets at the pharmacy, rather than waste lemons for a few drops of their juice? Or else blanch the vegetables, since the enzymes lose their capacity to react once they are heated? That's what cooks used to do, in fact, though often for odd reasons. Blanching, it was said, carried off the bitter molecules—a very mysterious culinary dictum.

CHLOROPHYLL ISN'T WHAT NEEDS TO BE FIXED

The question of what needs to be done to keep green beans bright green once they are cooked does not end with blanching. Far from it—cookbooks abound with theories in this regard. Many authors insist, in particular, that the chlorophyll has to be fixed.

Fix the chlorophyll? Let's examine these words closely, beginning with the last one. "Chlorophyll" does not exist. Instead there are several chlorophylls: chlorophyll a, b, c_1, c_2, and d. All these molecules have a similar chemical structure, which may be likened to a platelet (technically, a chlorin ring) with a long tail (a phytol chain). At the center of the platelet is a magnesium ion.

Prolonged heating of plant tissue containing chlorophyll molecules causes the magnesium to be eliminated and replaced by hydrogen. This is fatal to the lovely green color of green beans, for example, because chlorophylls in which the magnesium has been replaced by hydrogen become pheophytin molecules that give plant tissue a brownish yellow tint. In seeking to preserve the green color of vegetables, therefore we do not need to fix (or stabilize) the chlorophyll (though it should be chlorophylls), which remains in the plant tissue in any case, but to fix the *magnesium* in the chlorophyll molecules.

KEEPING THE GREEN BEANS GREEN

How can we put this knowledge to use—and keep green beans green? First, we need to use very fresh beans whose chlorophylls have not been degraded from sitting out in the produce section of the market for too long. Next, we have to be careful not to cook them for too long, lest the magnesium in the chlorophyll molecules be eliminated. Similarly, we must avoid at all costs cooking them in acidic water, for acidity is synonymous with hydrogen, the mortal enemy of magnesium. In earlier times cooks used an "ash detergent," made by filtering ashes and water, which yielded a potash solution—potash being a base that neutralizes acids and therefore protects chlorophyll molecules against magnesium loss.

Another classic method involved the use of a regreening tank, made of bare copper. Copper atoms detached from the surface of the tank replaced the magnesium in the chlorophyll molecules, imparting an almost fluorescent green to the vegetables—a dangerous technique, since copper is toxic. Fortunately, the practice of adding copper sulfate to the cooking water, which has the same effect, was outlawed in France in 1902.

Transmitting Knowledge

Marie-Odile Monchicourt: Conducting laboratory workshops, giving public lectures, teaching courses in molecular gastronomy—why are you doing all these things?

Hervé This: Because I have faith in rationality and in sharing. I am against keeping secrets. Imagine dying and taking to the grave with you the secret of where to find the best mushrooms. It is shameful to recall the story of the nineteenth-century French chef Joseph Favre. Observing on his arrival in Paris in the early 1880s that cooks had neither a professional journal nor a center for research, he took it upon himself to sponsor free public lectures and, in 1883, to found the Académie Culinaire de France. But when he created culinary classes for the layperson, he was reproached by his colleagues, who feared that their secrets would no longer be safe.

What is there to be gained by opposing attempts to contribute to the collective good? The aim of science is to add to knowledge and to disseminate it widely, so that a general and concerted search for technological applications will lead to the production of better designed and more useful food products.

Culture is what is held in common by the members of a group, and science is a part of culture. In an age such as ours, when philistinism reigns and money takes the place of moral value, knowledge is our strongest rampart against intolerance. Must it not therefore be made available to everyone? Must it not stand resolutely opposed to the seductions of *panem et circenses*—the bread and circuses of the Romans that we know in a different form today?

Of course, we must eat. But the laboratory workshops and various other initiatives I have undertaken must convince children that eating is not merely an animal act. Having discovered the wonder of culinary transformations at an early age, they will understand the importance of a sound and balanced diet, which in turn will enable them to travel the wonderful road to knowledge and freedom as they grow older.

In the process we must find ways to get across the idea that mathematical calculation can be fun, that science is one of the most entertaining games there is, and that technology is extraordinarily enjoyable as well.

M.-O. M: And what about cooking?

H. T.: In cooking there are all these dictums that cooks have accumulated over the centuries. Generally speaking, they are of two types: the correct and the incorrect. The correct ones are like wings that bear us aloft, that carry us to new heights. The incorrect ones, on the other hand, are like balls and chains that we drag behind us. This is why I began studying these rules in the first place, why I've been so intent on collecting and testing them. How can we in good conscience bequeath to our children false dictums that will only hinder their progress? Isn't raising a child a matter of raising her up above ourselves? This takes a great deal of work—as all parents know, children can weigh a lot. At bottom, this is my motivation in trying to transmit the results of molecular gastronomy to the next generation and to the ones that will come after it.

M.-O. M.: It's a question of love, then, isn't it?

H. T.: Yes, love—all the more since love is the spirit of cooking itself. I don't really know why I've spent so much time trying to understand it. At first I was interested only in the technical aspects of cooking; but gradually, as a result of having spent time in the company of talented chefs, I came to understand two things: that cooking is an art, and that art is more important than technique.

To make a light, airy soufflé is a fine thing, but if the soufflé isn't any good it doesn't matter. Which raises the question of what is good. The good is what is beautiful, and the beautiful—beautiful to eat, I mean, not to look at; we're not talking about painting or sculpture—is an aesthetic rather than a technical concept. By "aesthetic" I mean that it belongs to the branch of philosophy that concerns itself with the idea of beauty, not things that are beautiful to look at. And beauty cannot be produced by following recipes. I've thought about this problem a great deal with Pierre Gagnaire, whom I've given the nickname "the mad chef" (obviously this isn't an insult—I would be delighted if he were to call me "the mad chemist").

But to come back to my point about the soufflé, you have to give it flavor—and long before you put it in the oven. That's the main thing. I remember having a similar conversation with Christian Conticini, another very talented chef, about emulsions. He said he didn't care about stabilizing an emulsion if it didn't have the flavor he wanted in the first place.

In any case, going out to the restaurant of a culinary artist isn't a matter simply of going out "to eat"; it is more like going to a concert or to the opera, which is something one does for the soul and not the body.

This brings us back, finally, to the question of love. Serving a meal is to give happiness to others, not to supply nutriments: fats, proteins, carbohydrates, and so on. Even the best soufflé, both in nutritional and artistic terms, will be bad if you don't make your guests feel at home. A meal shared with disagreeable people, no matter how elaborate or well prepared it may be, will never be good either—whereas a sandwich shared with dear friends is a perfect delight. And our grandmothers, whose cooking we all adored, may not have been very good technicians, but what they gave us before everything else was love. Yes, cooking is first and foremost about love, and only then about art, and after that technique. This is such an important point that Pierre and I have devoted a book to it—probably my most important work to date, though much less technical than the previous ones.

M.-O. M.: The previous ones—in the very long list of your attempts to communicate your ideas, I'd forgotten to mention the books, articles, radio and television shows, and so on.

H. T.: Underlying all these things is a single conviction: that new knowledge is worth nothing if it is not disseminated far and wide. In the case of cooking, the most promising of my many projects is the foundation that has recently been created under the aegis of the French Academy of Sciences, the Foundation of Food Science and Culture. Here at last is the technical center that cooking has needed for so long. It is not restricted to members of the culinary profession; it is meant for the public as well, so its purpose is much broader than that of a purely professional society.

Of the foundation's six divisions, one—a technology division—will be responsible for revamping the way we cook. There is also an artistic division, which will concern itself with the question of culinary art. A public affairs division will be responsible for communicating the results of this research to everyone. (The other three divisions are science, hygiene/safety/regulation, and training.)

Molecular gastronomy is not at the center of this particular enterprise, having only a small place within the science division. And one of the foundation's missions, to come back to the question of communication, will be to try to reintroduce cooking in the schools—not in the form of mindless recipes, which transform the cook into someone who merely carries out instructions, but in the form of laboratory programs that encourage the search for knowledge. Science, its method and spirit, can be communicated to children. It's not a matter of naively supposing that all of them will become scientists, but of getting across the idea that the scientific method is essential.

And then there is the ecological dimension of cooking. As I mentioned before, a gas burner or electric coil wastes as much as 80 percent of the energy it consumes; the excess energy escapes into the atmosphere in the form of heat, twice a day in millions of households. This simply isn't right. We urgently need to change our behavior, to begin using better-designed kitchen equipment, ovens with more precise temperature regulation, induction coils, and so on.

Here again, we are faced with a massive task of communication. For me, this is a political act, requiring the most effective possible use of television, radio, books, magazines, and lectures.

Moderation in All Things

Marie-Odile Monchicourt: I recall your saying one day that you didn't like Blaise Pascal. This was a surprise, coming from someone who usually has such a positive attitude.

Hervé This: Yes—a mistake on my part, for one is happier when one speaks well of others than when one speaks ill of them. The reason for this outburst, which I regret, has to do with the boundless and wholehearted enthusiasm I feel for my work. I love it so much that I fight against anything that gets in the way. Besides, doing something that one loves before everything else can hardly be called work. I'm always having fun; in a sense, I'm always on vacation—and I get paid for it! To be sure, I have my share of administrative duties that distract me from my first love, which is doing science; I have my share of lecturing, teaching, and so on, which is sometimes exhausting. But, at bottom, my work is fascinating.

M.-O. M.: Fascinating?

H. T.: Perhaps I should use another word, since "fascinating" claims too much. Spinach, which some people like and others hate, is neither good nor bad. It is bad if you hate it, but good if you like it. My work is not fascinating in and of itself, or at least not everyone finds it so; but it fascinates me. I well understand that what is play for me may be an ordeal for others.

M.-O. M.: Is there a lesson in this for young people in school, pursuing their studies?

H. T.: Of course. I believe it is more valuable to study for the wonderful pleasures that learning gives us than to worry about getting good grades that will help us get a job. If certain subjects seem uninteresting, it's because we don't know how to appreciate them. We need to learn to seek knowledge for its own sake. Isn't knowing the same as loving?

To come back to Pascal. He made a distinction between the spirit of finesse and the spirit of geometry; or, to caricature slightly, between the humanities and the sciences. If I said that I didn't like Pascal, it was because I meant that I don't see why one can't like everything, the humanities *and* the sciences. Lavoisier, the father of modern chemistry, once said (borrowing an idea from the philosopher Condillac) that in order to do good science one has to be able to speak well, and vice versa. Why should children be allowed to say that they don't like math and science, or that they don't like art and literature? Do we want them to grow up one-legged? Shouldn't we take the view instead that it is a failure of teaching if they pursue only one line of study, turning their back on others?

It seems to me that one should specialize only grudgingly, while deliberately retaining a fondness for the paths one didn't take. I sometimes tell students in my courses on molecular gastronomy (which are open to the public and free at AgroParisTech, formerly the National Agronomic Institute) that every bad grade I give out is a punishment that I inflict on myself, because I haven't succeeded in making them work hard enough to earn a good grade. Obviously this is a bit of an exaggeration, but you get the idea.

M.-O. M.: And do you yourself like both the humanities and the sciences?

H. T.: Passionately—and for as long as I can remember. When I was in school I was always reading philosophy and literature and doing science experiments in my spare time. All my pocket money was spent on these two hobbies. I was a pretty good student, but I couldn't tear myself away from reading and experimenting, when I should have been paying attention to my other subjects. I wasn't very disciplined—I couldn't keep my mind from wandering. I spent a lot of time daydreaming in class.

It all began when I was about six and I received a chemistry set. My father also told me that I could borrow any book in his library as soon as I knew how to read. I quickly got in the habit of devouring every book I could lay my hands on, while devoting the rest of my free time to doing experiments, even if it meant falling behind on homework and other obligations.

Later, when I was studying chemistry at the École Supérieure de Physique et Chimie Industrielles de Paris, I enrolled in modern literature courses on the side. I couldn't see why I should deprive myself of the pleasures of literature just because I was training to be a chemist. Don't we have the right to like everything that deserves to be liked?

Steak and French Fries

With the leg of lamb we examined braising, a technique perfectly suited to relatively tough cuts of meat. It remains to consider the case of tenderer meats that are suitable for grilling. To make matters simple, let's consider steak. What should we serve with it? Most people like starches. French fries are the obvious choice, all the more since their mode of preparation—frying—is quite interesting from the physical and chemical points of view.

Constructing a Dish

Marie-Odile Monchicourt: In the last chapter we considered a leg of lamb, whose texture we constructed, accompanied by green beans, whose color we constructed. With steak and French fries we carry on with simple cooking, but what are we constructing in this case? Aren't we simply trotting out another old warhorse of classical cuisine, without the least hint of originality?

Hervé This: Steak frites is a wonderful dish, precisely because there's nothing remarkable about it. We are forever demanding something out of the ordinary—as if we wanted music always to be played very loud. This is a mistake, for we perceive forcefulness only in alternation with gentleness. Our sensations depend primarily on contrasts. And innovation can only come about through opposition.

That said, steak and French fries can be interpreted in quite a number of ways. The fries can be more or less thick, more or less regular, more or less crispy, and so on. In good cooking, everything is thought out in advance. Nothing should be left to chance. Steak and fries also raise the question of love: How should these two elements be presented so that the diner sees at a glance that someone has taken trouble on her account—that she is loved? Can we be satisfied with haphazardly throwing the fries on a plate, or are we better advised to give some thought to their arrangement?

Organizing and constructing—these are the two watchwords of the new tendency that I call culinary constructivism, and that I try to promote at every opportunity.

M.-O. M.: What of molecular cuisine, which is much also talked about these days?

H. T.: It's very fashionable at the moment, but it will only be a transitional phase. It involves introducing new utensils, new ingredients, and new methods into cooking. The idea has taken hold slowly, but results can already be seen. Cooks today almost routinely use gelling agents extracted from algae, liquid nitrogen to make sherbets and ice creams; they distill, and do many other novel things as well.

Molecular cuisine is a revitalized form of cooking; it's not content to go on repeating the recipes in Escoffier's *Guide Culinaire*, which was published more than a century ago now and has influenced generations of cooks. Even so, a tendency—a fashion, if you like—that is ten years old is already outmoded for true innovators. This is why a few years ago I began to think about the next fashion, the style that could most usefully follow molecular cuisine. The most interesting thing today, I believe, is not abstract cooking, which I introduced not long ago, nor note-by-note cooking, which I proposed after that, but culinary constructivism, which seeks to construct dishes by taking into account every aspect of their preparation and presentation, every possible sensation.

The Question of Steak

To make a good steak, it is said, the meat must be good. With the leg of lamb we saw how tough meat can be tenderized. What we need to do now is to think about the cooking of meat in general, without reference to the precise characteristics of this or that piece of meat.

Let's begin at the beginning, then. Meat is made of muscle fibers bound together by collagenous tissue. Either the meat is tender, because it contains little collagen, or it is tough, because it contains a lot of collagen. This is somewhat misleading, because there are different kinds of collagen, but let's keep things simple for the moment. When meat is grilled it becomes tougher, because the proteins inside the fibers coagulate. When it is braised it becomes tougher as well, but the principal phenomenon, as we saw with the leg of lamb, is a tenderizing of the collagenous tissue. In other words, the chief consequence of grilling a steak is that the meat becomes tougher.

WHAT IS COOKING?

Cooking does several things. Lovers of raw meat must be reminded that cooking kills the pathogenic microorganisms that are typically found on the surface of foods and, in some cases, kills parasites that contaminate the inside of foods (one thinks of pork, horsemeat, and the flesh of certain fish). It also has the effect of modifying the texture of foods and, finally, of imparting flavor to them. All these things occur simultaneously during an operation as simple as preparing a steak.

It is often said that in order for a steak to be properly cooked, it must be seared because the crust that forms on the surface prevents the juices from escaping. True or false? The reasoning is false, as it turns out, but the conclusion is true. Let's compare two pieces of the same meat, one of which is quickly seared and the other slowly cooked. In the latter case, gradual heating causes the collagen to contract, with the result that the juices run out and thereafter slowly evaporate. In effect, because the evaporation of these juices limits the surface temperature to 100°C (212°F), the meat boils in its juices.

In the contrary case, when the meat is seared, a thin crust is formed by the rapid evaporation of the water on its surface. A thermometer placed beneath the surface of the meat shows that the temperature rises considerably, far above the boiling point. This is why the surface browns, as a result of various chemical reactions: oxidation, hydrolysis, Maillard reactions, and so on. This browning is accompanied by the formation of a great many odorant and taste molecules, as in the case of coffee or chocolate that is roasted. And since the browning occurs rapidly, the actual cooking of the meat doesn't last long, which limits the amount of juice that escapes: The meat remains tender because the inside is rare. Classical cuisine was therefore right to insist that meat be seared as quickly as possible, but not for the right reasons.

If further proof is wanted that the crust does not retain the meat's juices, we have only to observe a piece of meat as it cooks. Rising from the meat we will see steam, which is to say the water that has come out of the meat (a sign that the crust is not impermeable) and that subsequently evaporates. Another sign, of course, is that when the steak is cooked and then transferred to a plate, it quickly finds itself bathing in a sea of juices—which have continued to come out of the meat.

Should Meat Be Salted Before, During, or After Cooking?

Marie-Odile Monchicourt: You concern yourself with cooking down to the smallest details. Have you gone so far as to consider when grilled meat should be salted?

Hervé This: Even further—I've made it the theme of one of my monthly INRA molecular gastronomy seminars. Each of these sessions, which bring together chefs and scientists (as well as journalists and recreational cooks), poses a question, the answer to which is to be determined by experimentation. The seminars themselves are spent discussing details of experimental procedure and thinking up new experiments, the results of which are reported at later sessions.

During one of the first seminars, the question came up of when to salt grilled meats. The professional cooks were divided into three groups; those who salted before cooking, in order—they said—to allow the salt to penetrate the meat; those who salted during cooking, in order—they said—to give the meat a golden brown appearance; and those who salted after the cooking was done, in the belief that the salt would make the meat "sweat," causing it to release its juices.

But why, it may be asked, should the meat be salted in the first place? The question seems almost silly. One salts a piece of meat so that it will be salty, right? In reality, however, the matter is far too involved to be disposed of with a snappy reply of this sort. All cooks know that an unsalted vegetable soup, even one made with good vegetables, has no flavor. Put a little salt in it, however, and all of a sudden it takes on flavor—the flavor of the vegetables. What happens is that the salt, in binding itself to the soup's water molecules, triggers the release of volatile molecules that have little affinity for water—among them odorant molecules, which readily detach themselves from the liquid and, passing into the air, eventually reach the olfactory receptors of the nose. In other words, salting a soup gives it more fragrance, which is perceived in the course of eating as the odorant molecules rise up into the nose through the retronasal fossae at the rear of the mouth.

M.-O. M.: And did you resolve the question of salting the steak during the seminar?

H. T.: Not during the seminar, but afterward. The seminars are devoted mainly to encouraging everyone to experiment and to discussing protocols; the experiments themselves are usually carried out on the side, because they take a lot of time. To determine whether salting causes the meat to sweat, I coated a steak with salt and let it sit for several weeks, weighing the meat every day. It turned out that different cuts of meat reacted very differently. With steaks that are cut along an axis parallel to the fibers you notice on the surface—hanger steak, skirt steak, and so on—salt brings out very little juice. With sirloins and other rib steaks, however, which are cut perpendicular to the fibers, the juice runs out very quickly. The same is true for chicken breasts, for example.

Obviously such sweating experiments are inadequate by themselves to answer the question. My colleagues and I also used electronic microscopy, in combination with chemical analysis, and discovered among other things that salt does not in fact penetrate steaks on the grill. In the end, it turned out that none of the three opinions under examination was any more justified than the others—so you should feel free to salt as you please. When you really get down to it, there is no right way. In cooking, the good is what *you* like.

HOW MUCH SALT SHOULD BE USED IN SEASONING MEAT?

The general question of seasoning is extremely important in cooking. In the case of salt, the problem seems quite intractable if one takes a narrow view. Each person likes salt in his own way; some like their food with only a little, some with rather a lot. One has the impression that salt is like a volume control on a stereo system that can be adjusted over a scale of 1 to 10. Is it possible to satisfy everyone?

A culinary artist like Pierre Gagnaire replies that cooks who work with the idea of a volume dial or cursor in mind will succeed in pleasing only those who like salt as much as they do. It is more useful to think of salt not as something that is used because seasoning food is the only technique in the chef's repertoire, but because seasoning is but one of many instruments in an entire orchestra of flavor that he conducts. Just as a composer scores the melody of a clarinet or of violins in a work, a chef incorporates salt in his own orchestral composition, where seasoning occupies only one chair among many. Obviously, this assumes that the chef has an overall view of the composition, but can it really be otherwise? I don't think so.

ANOTHER WAY OF COOKING STEAK

Grilling is not the only way to cook steak. It can also be panfried in fat. This amounts to a kind of braising, only with the difference that the meat has initially been seared over high heat, and so has a crust. There is also, of course, the old-fashioned method of studding meats with small pieces of bacon.

The practice of larding meat, like that of wrapping quails with a thin slice of bacon before roasting them, turns out to have been correct from the chemical point of view. In the 1990s the British chemist David Mottram discovered that Maillard reactions produce different odorant compounds depending on the presence or absence of fat. Bringing out the flavor of the meat, therefore, depends on how much fat the meat contains. (If you cook your meat in fat and you want to keep your figure, you'd better wipe off the grease.) That is the point of studding meats with strips of bacon. And if care is taken to season the bacon with salt and pepper beforehand, the meat will be especially flavorful.

Science and Diet

The work of the molecular gastronome consists of observing, measuring, noting, evaluating, in short, in understanding—for example, whether the meat of a pot-au-feu should be put in boiling water (as some recipes indicate) or whether the cooking should begin with cold water (as others direct). Molecular gastronomy therefore has nothing to do with dietetics, which has a behavioral (or "moral") component; science, by contrast, limits itself to exploring phenomena.

Diet, it should be noted in passing, is a fertile source of prejudices and contradictions. Many of the same people who object to the least residue of pesticides in their fruits and vegetables do not think twice about cooking fats at such high temperatures that acrolein—a molecule that is very harmful to human health—is produced. Many of the same people who insist upon eating healthy foods smoke tobacco. Many of the same people who want to lose weight refuse to exercise. I mention these things not as criticism, only as observation. We often eat in an irrational way because, as we have seen, eating is a matter of culture. Chocolate is composed principally of fats and sugar; but if it is in our culture, we eat it—and then complain about being overweight.

When it comes to eating, one may well wonder whether it is wise to eat potatoes or bread or meat or fish—or French fries. But questions such as these are of no concern to molecular gastronomy, which seeks only to understand what happens to food when it is cooked. Later it will be necessary to find ways to apply this knowledge, particularly with regard to its dietary implications.

Good for Your Health?

Marie-Odile Monchicourt: Do we eat worse today than we used to?

Hervé This: No one can say for sure, but I am inclined to doubt it, because we know that life expectancy continues to increase, in industrialized countries. But what does eating badly really amount to? Does it mean eating things that are bad for one's health, or merely things that are thought to be objectionable for one reason or another? A great deal of nonsense is spoken on the subject of food in relation to health. We should be wary whenever we hear the word "natural" or the phrase "good for your health." They usually conceal a vested commercial interest or an ideological agenda of some sort.

The main point that needs to be made is that we don't behave rationally when it comes to diet. You will recall the episode a few years ago in France, when chickens imported from Belgium were found to have illegally high levels of dioxin—the whole country was up in arms. And yet no one makes a fuss about the fact that, when we eat grilled meats, we consume considerable quantities of benzopyrenes, which is to say carcinogenic molecules deposited by the smoke. Why do we get upset about the presence of dioxin in chickens while we merrily go on grilling meats in a dangerous way? Our ancestors knew that meat should be put in front of the fire and not on top, where it is unavoidably licked by the flames, leaving a black deposit of sparks and benzopyrenes. Placing the meat in front of the fire (where the infrared rays will cook it just as well) has the further advantage that we can put a pan underneath to catch the delicious juices that drip from it—and, behind the meat, a semicylindrical piece of metal (known as a *shell*) that accelerates the cooking by acting as a heat reflector. To cook well, you have to think about what you're doing.

M.-O. M.: But for every problem that's solved, another one comes along to take its place. Many foods are contaminated, for example.

H. T.: Contaminated? It is often the word that frightens us, distracting our attention from the facts of the matter. The reality in this case is that in industrialized countries we are very well protected by government agencies, particularly the departments responsible for detecting and prosecuting commercial fraud.

French food and drug officials attend the INRA molecular gastronomy seminars, by the way, and listen to chefs' complaints while explaining the reasons for the regulations they are required to enforce.

Concerns over toxicity levels in food are, I believe, misplaced for the most part. Every cook is in the habit of using spices and aromatic herbs, such as nutmeg, tarragon, and bay leaf, without the least sense of unease, because few of us are aware that they contain frighteningly toxic molecules (trimyristin, estragol, and so on). A nutmeg berry ground into powder can kill a person. The question of dosage, or proportion, is crucial. It is therefore meaningless to use phrases like "good for you" and "bad for you" without specifying the quantities involved. For example, alpha-amanitin, the chief toxic agent in the lethal death-cup fungus (*Amanita phalloides*), is also present in the chanterelle mushroom in very small amounts. Should we therefore stop eating chanterelles?

Spices and aromatic herbs have powerful flavors because they contain high concentrations of odorant and taste molecules. We must therefore follow Paracelsus's principle—"Nothing is poison. Everything is poison. The difference is in the dose."—and use them in very limited amounts: one or two bay leaves, a little grated nutmeg, and so on.

M.-O. M.: "Natural cooking" is much talked about right now. What is your view?

H. T.: Natural cooking? There can be no such thing, by definition. To cook, by its very nature, is a form of technical intervention. Cooks transform nature, just as chemists transform matter. To be truly natural we would have to serve only raw foods—foods that for the most part are hard, contaminated, and inedible. No, cooking is the acme, the epitome of the artificial. Consider that a cook thinks nothing of heating oil to 200°C (almost 400°F) in order to make French fries; a chemist, working in a laboratory well protected by fume cupboards, fire extinguishers, and so on, only very rarely deals with such temperatures. And in roasting a piece of meat or making caramel, we set in motion very vigorous chemical reactions. There isn't anything at all natural about cooking—and it can't possibly be otherwise. Why do we think that cooking ought to be natural? This is another of our fantasies that it would be interesting to examine more closely.

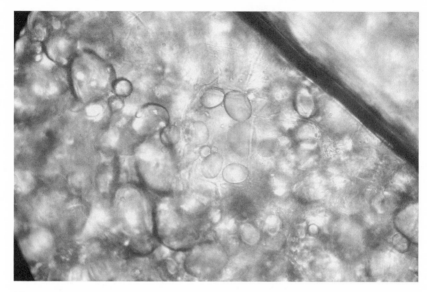

Microscopic picture of a raw potato. Potato cells are small edged sacs that enclose starch granules (the round forms).

Preparing French Fries

This is not the place to decide whether French fries are good for us or not. Our job is to make the best French fries that we can. Whereas meat is composed of muscle fibers, which is to say elongated cells, the cells of potatoes (like other vegetables) are more rounded; in both cases they contain water and everything else that cells need to live. The tubercular cells in potatoes stock reserve supplies of energy in chemical form. More precisely, they contain starch granules, which most every child (and every adult) has seen at least once under a microscope. But you don't need a microscope to see them. You have only to cut up a potato and put the slices in water. The white particles you see that make the water cloudy are starch granules. And if you grate the potato, the granules form a powdery sediment that resembles flour, like Maizena (a brand of corn starch sold in many countries) or potato starch (which is what it is).

This is why you have to wash the potatoes after they have been sliced. If you don't, the starch granules that are released in the process become dispersed in the frying oil, where they burn and turn the oil dark. The potato sticks should also be dried after they have been washed, because otherwise the water that adheres to the surface will evaporate at the outset of cooking, lowering the temperature of the oil. And if the oil isn't hot enough, you will end up with greasy fries.

THE PRINCIPLE OF FRYING

The potato sticks are cooked in oil that has been heated to a very high temperature, usually 180°C (about 350°F). The surface moisture evaporates at once (hence the bubbling up of the oil), yielding a dry outer layer of cells to which the starch granules are sealed in the form of a crust.

Why should a very high temperature be necessary for frying? If you hold a potato stick by one end and immerse the other one in 180°C oil, you will not be burned—proof that a potato is a poor conductor of heat. This is why the oil must not be too hot to begin with: otherwise the surface will burn before the inside has time to cook. But if the temperature is too low, the oil will readily penetrate the surface and the fries will be greasy. What to do?

Remove a stick from the oil and you will see that the water it has lost, under the crust, has been replaced by steam, which, because it was under pressure, escaped through the holes in the crust. When the stick cools down, the oil coating its surface will be absorbed as the steam inside recondenses. If we measure the pressure in an individual stick, we will observe first an increase at the outset of frying; then, after the stick has been removed from the fryer, a minute elapses before the pressure begins to diminish. In other words, you have a minute to wipe off the French fries before they absorb the oil that coats their surface. In this way three-quarters of the oil can be prevented from seeping inside. The result: less greasy and better fries. This principle holds for all fried foods.

Lemon Meringue Pie

Dessert poses still more questions for those who know how to see them. Why should we serve something sweet at the end of the meal, something whose taste is sure to please children—so much so that dessert is what we deny them when they are naughty?

A lemon meringue pie is an obvious choice for dessert because it consists of a series of layers, like the pies you find in pastry shops and restaurants, but it is much simpler. Before we go into the kitchen, however, let's leave our own country for a moment and travel abroad, where it is easier to see the beam in our own eye.

In a restaurant in Greece, the waiters bring out the food on trays, and then the chef comes out of the kitchen to add to each guest's plate a spoonful of sauce, or a dollop of cream, or a dash of spices, or what have you. In another restaurant in another country, a meal is placed in the middle of the table to be shared among the commensals (from the Latin *com* [together with] and *mensa* [table]); everyone is happy until the time comes to decide who gets the cherry on the cake. In still another country, the door to the restaurant is shut and one has to ring the bell in order to be let in. Finally, returning to Greece, in a fourth restaurant—a run-down place frequented by elderly people who feel more at home there than at the trendy café next door—the chef comes out to personally serve each person his meal, with a word or two for each of them, particularly an old woman, alone and sick, with whom he stays for a time, even cutting up the meat on her plate for her.

In these four places we will find the keys to a successful meal, because they illuminate what I mean by love. Why should this component of cooking—the giving of happiness—be important? Biologists have shown that the phenomena of the living world must be interpreted in the terms of evolution, for it is natural selection, acting in combination with spontaneous mutations, that has made the human species what it is today. In particular, selection has caused our species to be sociable. Like other primates, we live in groups, and exclusion is perceived as a horrible thing, precisely because our genes have selected for this evolutionarily advantageous behavior, which, by preventing us from being isolated, favors survival and the reproduction of the species. Sociability is what leads us to smoke tobacco (if we don't we risk being rejected by those who do) and to drink beer (despite its bitterness, for if we don't drink it we are excluded from the society of drinkers). Gathering to eat can likewise be seen as a kind of biological reward, elaborated over millions of years of evolution. Eating the cooking of others is a way of uniting ourselves with them—to the point of entrusting them with our lives.

On Love in Cooking

Consider what is happening in the first restaurant. What does it mean to add a little something to each person's plate? To be sure, a team of cooks in the kitchen has prepared each guest's meal, taking care over its presentation and so on, but the chef who comes out into the dining room and personally adds a touch of this or that is saying to his guests, in effect: "I love you so much that I can't bear to see your plates leave the kitchen without coming out myself to add a dash of love to them."

In doing this the chef is acting spontaneously, but he is also showing that he understands the essence of his vocation. His gesture reminds us of the cooks in our school cafeterias who personally serve the children, one by one, adjusting the amounts on each plate as needed, with a word for each child and a smile, giving big eaters a bit more, encouraging those with smaller appetites, adding a pinch of parsley here, a pinch of salt and pepper there. The children are grateful for such attention. Instead of receiving merely a plate of food, each one receives love. It doesn't matter whether their plate contains fish or meat: They are about to receive something more.

In the second restaurant, where the guests share a meal, there is true pleasure, because they enjoy a sense of companionship. They share not only a table, but their food as well. Drinking out of another person's glass, it is said, allows us to know his thoughts. The same is true of being served from the same plate—one has the illusion of knowing the flavors that another person has in his mouth. I say "illusion" because, as sensory neurophysiologists have demonstrated, individuals detect odors and flavors differently. In particular, the detection threshold for sweetness, which is to say the smallest quantity of sugar perceptible in a given volume of water, varies from person to person, which explains why we put different amounts of sugar in our coffee, for example.

Nonetheless, in the sharing of a meal, there is an exchange of mutual affection—assuming, of course, that our companions do not spoil the magic of the moment by quarreling over who will get the cherry on the cake. Unless the guests have decided in advance, explicitly or otherwise, that the cherry is to go to one of them in particular, the presence of a single cherry on the cake must be considered the fault of the pastry chef—because it cannot be shared.

And how do we prevent some from taking more than others? In many cultures, the solution lies in excess. Dishes such as choucroute, cassoulet, couscous, and so forth are prepared in such ample quantities that there's enough for everyone. Nonetheless there is always the chance that for some of the guests a sense of injustice will ruin the pleasure of sharing. One would have liked to have a certain piece of meat, which another took for himself, or indeed the cherry—the only cherry—on the cake. The chef could resolve the problem by placing as many cherries on the cake as there are guests, of course, but in that case the element of sharing would be lost. Can a way of sharing be devised that everyone will be happy with?

Before replying to this question, let's consider the restaurant whose front door is closed, whose guests are left waiting outside for someone to come open it. Can anything worse be imagined? We expect a restaurant to welcome us; we expect the chef to assume responsibility for our well-being for as long as we are under his roof, to paraphrase Brillat-Savarin's famous aphorism. The chef who wishes to show us love must welcome us, as Daniel Vézina has done at the entrance to his restaurant in Quebec City by creating a sort of vestibule filled with baskets of apples. In this small enclosed space, the apples release an enchanting

The presentation of a dish is a way of saying "I love you, for you can see that I have constructed this dish for you."

fragrance that invites us to enter: What a delight! The shut and locked door, by contrast, announces a culinary morgue: The chef is not interested in showing us love; he wishes instead to exert control over us, to enlist us as witnesses to his glory, to subjugate us. We should never tolerate such treatment, which dishonors the profession. We must denounce it—and move on at once to another destination.

The restaurant that has seen better days, where the chef, no longer young himself, unobtrusively comes out from the kitchen to serve each diner personally and to say a cheering word to those in need of one, is a profoundly human place. What is most striking is the contrast with the trendy café next door. There one finds a waitress in a miniskirt, customers talking on their cell phones, designer furniture, loud pop music, bright lights; here, a carafe of water for each person, marble tables covered with oilcloth and chairs that can be moved around to accommodate larger parties, or separated when guests wish to enjoy a bit of privacy. Naturally the cooking in the old restaurant is traditional; better still, the traditional dishes have not been deconstructed. The salad is perfect, not because of any affectation, but because of the care taken in cutting the onions in such a way that their crunchiness harmonizes with the tomatoes, which themselves have been sliced just so, and in garnishing the composition with a thick slice of feta cheese, not omitting the quartered lemon and scented olives. The chef has not sought to transform the ingredients in complex ways; instead he has taken the trouble to produce a familiar dish simply and well. The other dishes show the same degree of attentiveness, which perfectly reflects the manner and character of a kind man who goes out of his way to make everyone feel at home.

All of this inevitably reminds one again of the essential role played by cooks in school cafeterias. Let us recall with gratitude—and fondness—all those who, day after day, concern themselves with the well-being of the children they feed. "Oh dear, you don't look quite right—a bad grade? How did that happen? Come on, have a little more sauce, that will fix you up. Tomorrow you'll do better. And you there, how are you? You look hale and hearty. And you, oh my, your mother is still sick—eat a good plate, my dear, that will build up your strength. . . ." And so on. All these saintly people who cook in prisons, hospitals, and schools deserve our thanks.

What Do Humans Eat?

This question may seem to suggest that all human beings eat the same thing—whereas it is plain that there are innumerable differences among individuals, whether young or old. Nonetheless we will have a better understanding of our dietary behavior if we examine the behavior of our fellow primates, while taking care, of course, not to fall into the reductionist trap of supposing that the development of the individual recapitulates that of the species.

Studies of the feeding habits of primates have shown that both chimpanzee and human newborns grimace when a sour or bitter substance is put on their lips, but display a common expression of pleasure when they are given something sweet to taste. Among the primates, human and other, evolution has "coded for" the pleasure that accompanies the consumption of sweet foods and liquids—sweetness being associated with energy, which is indispensable to life. Conversely, bitterness seems to be associated with the toxic alkaloids found in plants.

These studies show that the human species is less exceptional than we like to think. The majority of nonhuman primates feed on fruits (strictly speaking, the edible pulp of phanerogamic plants), which contain variable concentrations of soluble sugars (chiefly fructose, glucose, and sucrose). This preference is activated by an ability to recognize the taste of these sugars. Measuring the perceptual thresholds for sweet substances, we discover an "allometric" relationship, namely that the attraction to sugars varies over species or, more precisely, that gustatory acuity with respect to sucrose and glucose increases with body mass (thus the largest species have the lowest perceptual thresholds). This relationship is probably explained by the fact that small primates require relatively little energy, whereas large animals, which must supplement their diet with fruits containing relatively few sugars, need to be able to efficiently detect the presence of such substances.

Our species unconditionally confirms the allometric relationship observed in primates—unsurprisingly, perhaps, since we are ourselves primates; but there are nonetheless a great many differences among various human groups. Sensory recognition of soluble sugars among some peoples living in tropical forests, for example, is less acute than that of peoples living in more open environments.

Pies, Fats, and Sugars

After this long detour we arrive at our lemon meringue pie. The pastry is made of flour, butter, and sugar—enough to satisfy the primate in us. Sugar, as we have just seen, is biologically synonymous with energy and so indispensable to our survival. A taste for starches, as for fats, was thought until recently to be the result of conditioning, and in fact our system breaks down starches fairly slowly, which means that the glucose they contain circulates in the bloodstream for several hours.

A dog that is fed each time a bell is rung ends up salivating when it hears the bell. In the same way, we learn to consider starches, which confer lasting physiological benefit, as being good for us. The same goes for fats, whose odorant molecules have to be dissolved in order to be associated with the perception of flavors (our body contains specific receptors for these molecules—a recent discovery).

In other words, a pie pastry is a dietary *summum* from the biological point of view. And yet simply throwing together a mass of flour, butter, and sugar will not give you a pie pastry. Once again we see that the work of the cook is mainly a matter of knowing how to organize ingredients.

PUFF PASTRY

There are a great many kinds of pastry—what physicists call solid suspensions, for reasons that we will examine later. The classic types of shortcrust pastry are *pâte brisée* (basic pie dough made with flour, water, and butter) and *pâte sablé* (the same dough enriched with egg and sugar—hence its English name, rich shortcrust or sugarcrust pastry). And then there is puff pastry, which is rather more difficult to make. Let's begin with that.

"Roasted flour" is easy to make (you brown the flour under a broiler) and perfect for certain purposes, such as making cookies, where the gluten proteins can cause problems. Roasting deprives these proteins of their ability to harden the cookies.

Puff pastry is made by kneading flour with water in order to make a dough, which you then stretch to form a fairly thick square. Next you put some softened butter in the middle of the square, and fold the dough as if you were making an envelope. At this stage, there is one layer of butter between two layers of dough. After cooling the folded dough square in the refrigerator, you roll it out until it is about three times as long as it is wide, and then fold it in thirds. There are now three layers of butter and four layers of dough. This operation is then repeated five more times. Let's count the number of layers of butter: 9 after the second folding, then 27, 81, 243, and finally 729. The number of layers of dough? One more than the number of layers of butter. The final result is only about a centimeter thick.

The water in the dough evaporates during cooking. A very small quantity of water produces a large volume of steam (a gram of water, for example, produces about a liter of steam), which, by puffing up the dough and separating its many layers, allows them to dry. At the same time various chemical reactions occur that cause the dough to brown—and eventually you wind up with puff pastry, the flakes of which are less than a thousandth of a centimeter thick. It is easy to see why pastry chefs insist that the dough be worked in as uniform a fashion as possible: If it is not very carefully rolled out in between foldings (a process known as *tourage*), there is a chance that adjacent layers will stick together in one or more places, preventing them from separating. The *tourage* of the best chefs is so regular and so precise that the layers rise perfectly, forming extraordinarily delicate and airy flakes of pastry.

SHORTCRUST PASTRY

Until 1950 one spoke only of *pâte à foncer* (the basic dough used to line a pie pan) and *pâte sucrée* (when sugar was added to the dough). The fact that pastry crusts, once cooked, are more or less crumbly suggested another way of distinguishing between them. Thus crumbly crusts came to be called *pâtes sablées* (sugarcrust pastry) and ones that break without crumbling *pâtes brisées* (literally, broken-crust pastry).

This difference in friability can be traced to an experiment performed in the 1750s by the Alsatian chemist Johann Meyer (1705–65) and also by the Italian chemist Giacomo Bartolomeo Beccari (1682–1766), who was studying the composition of flour. They kneaded flour and water until it formed a dough that became harder the longer it was kneaded. Immersing the ball of dough in water and gently kneading it a while longer released a white powder, leaving only a yellowish substance having the consistency of chewing gum. The white powder we now call starch, and the yellowish remainder, gluten.

This starch, the same substance found in potatoes, rice, and so on, consists of small round granules made of complex sugar molecules. Gluten is comprised of proteins that are organized in the form of a lattice. Dough made from flour and water may therefore be thought of as a sort of net for catching fish, in this case starch granules. This is why pie dough is technically considered a solid suspension. Solid granules (starch) are dispersed in a solid continuous "phase" (gluten).

Cooking this dough yields a hard crust, since the granules are trapped and held in place by the network. By contrast, if we were to knead the flour first with butter instead of water, the flour granules would be coated by the butter, with the result that if they are then mixed together with water, they do not form the gluten network needed to harden the dough. The butter serves as a sort of soft cement, loosely binding the flour granules to produce shortcrust pastries of various kinds.

Another simple experiment shows that the sugar in sweetened pastry dough is an additional source of friability. First, make a dough from flour and water. Then, once the dough coheres in the shape of a ball, add some powdered sugar. With further kneading, the sugar comes to be incorporated into the dough, where it traps the water more effectively than the gluten—and the ball of dough collapses. After cooking the crust is crumbly, because the gluten networks have been destabilized by the sugar.

HOW MUCH BUTTER AND FLOUR ARE NEEDED?

General principles are useful, but to make a good pie dough we need precise measures. How much flour, butter, and water are required? The basic idea is to disperse small granules of flour in some butter. An elementary calculation yields the proportion of one part flour for one part butter. Nonetheless, if we stop to consider that the dough is made of granules of flour dispersed in butter, we will see that this can vary within very broad limits. This is why we can go so far as to use one part butter for thirty parts flour and still obtain a dough—though it will more resemble the dough used to make chapati, a flat Indian pastry, than the one we are accustomed to use for making pie pastry. The moral of the story is that the recipes give only limited guidance. It's better to understand what one is doing and then to decide, in perfect liberty, what one wishes to do.

For example, does a crumbly crust pastry have to have sugar in it? No, but the addition of sugar will prevent a gluten network from forming; at the same time it caramelizes as it cooks, imparting a delicious flavor. As for the amount of

In a ball of dough made from flour and water (*left*), the flour granules are trapped in a gluten network consisting of proteins bound together by the water. The addition of sugar (*right*) causes the structure to collapse since the sugar molecules trap more water than the proteins.

water that should be used, keep in mind that the water can be replaced by any tasty solution: an infusion of licorice, strawberry juice, wine, whisky, you name it—so long as it's got water in it. Nor should salt be neglected, for we know (though we do not yet know why) that it works to mute disagreeable flavors and to enhance more agreeable ones.

The Lemon Custard Filling

Imagine that you are a Martian chemist visiting Earth and that you observe Earthlings enjoying a lemon pie. On your return to Mars you wish to replicate the dish, but you forgot to write down the recipe. What do you do?

You can re-create it if you realize that the custard filling will have to have the taste of lemon, while remaining soft once it has set. Such preparations are known as gels—the result of proteins that have gelled, either physically or chemically. Egg proteins, for example, cause a juice to coagulate. This is the basis for the class of thickened sauces represented by custards. So if you add an egg to sweetened lemon juice and heat the mixture, you will obtain the desired filling. How much lemon and how much egg? As a chemist who understands the structure of gels, you can easily calculate that the minimum quantity of gelling agent needed, as a proportion of the total mass, is about 1 percent.

In other words, an egg weighing 60 grams (2 oz), with its 8 grams (.28 oz) of proteins, will yield 800 grams (28 oz) of custard. Obviously this is only a rough order of magnitude, and in practice it may be difficult to obtain more than 700 grams (25 oz), but the calculation has the advantage of showing that the smaller the quantity of egg, the softer the custard.

COOKING THE FILLING

Modern ovens incorporate a number of different functions, coming equipped with a convection compartment, a broiler, multilevel heating chambers (allowing dishes to be cooked on the floor of the oven, for example), and so on. Since an egg coagulates at 61°C (142°F), cooking by convection at 200°C (392°F) long enough to caramelize the sugar in the pie dough runs the risk that the custard will have dried out. A better way to proceed, then, is to place the dough and filling in a metal mold, which will efficiently conduct the heat to the dough, and cook the preparation on the floor of the oven; or else to cook the dough by itself, and then to cook it a second time with the filling just long enough for the custard to set; or else ... There are any number of possibilities for those who are willing to think for themselves, rather than simply follow a recipe.

The Meringue

Making a lemon meringue pie involves doing several different things at once: cooking the sugarcrust pastry at one temperature, coagulating the egg at another temperature, and heating the meringue at a third temperature. We need to consider the complications introduced by this third element.

What is a meringue? Egg white with some sugar added. A meringue in the French style has a hard exterior crust protecting a honeycombed interior, and as in the case of a Norwegian omelet, there is a danger that it will create an isolating layer that prevents the filling from cooking. As it happens, this style of meringue is not the only option available to cooks. An Italian meringue, for example, which is made by stiffly beating egg whites and combining them with a sugar syrup heated to a temperature of 120°C (248°F), is softer than the French version and can be placed over the cooked filling. One might think of putting the fully assembled pie under the broiler for a moment so that a hard layer is formed on top, without overcooking the filling, which will be protected by the meringue. (Foam is a good insulating material—think of the spray foams used in construction or, in solid form, the bubblewrap used for packaging fragile items.)

In this connection we should pause to take notice of recent research on the physical properties of crunchiness, which shows how apparently unrelated studies are apt to converge when you keep thinking about something long enough, as Pasteur used to say. An unexpectedly fruitful approach has been to model crunchiness in terms of the physical theory of percolation. True percolation—the kind you see in French coffee presses, for example—involves applying downward pressure on a volume of water so that it passes through ground and packed coffee granules, from top to bottom. The water makes its way through the granules following every path imaginable—and winds up producing a drop of coffee. This is the beginning, or threshold, of percolation.

A phenomenon of the same type occurs when one passes an electric current through a metal grid or latticework. If by chance one of the cross members is severed, the current takes a more complicated route than the original one, but it continues to circulate—until there is no longer a continuous path connecting one side of the grid to the other. At this point a percolation threshold has once again been reached. Something similar happens in a population when a virus propagates from one individual to another: Once the virus is sufficiently contagious, the country is sick from one end to the other. This is an epidemic, which likewise starts from a percolation threshold.

Researchers studying crunchy materials, such as meringues, hypothesized that this physical state derives from a propagation of microscopic cracks. Crack propagation does not occur in a soft material, however, which can be demonstrated by dripping syrups that have different degrees of water content over cold slabs of marble. Syrups heated to less than 127°C (261°F) remain soft, because enough water remains to prevent the sugar molecules from aggregating, whereas syrups heated to higher temperatures are crunchy because there is no longer enough water to prevent the formation of a continuous sugar network.

In the case of an Italian meringue one is liable to encounter exactly this phenomenon, because if the syrup is heated to too high a temperature, it vitrifies as it is folded into the stiffly beaten egg white. The percolation threshold of 127°C is therefore something pastry chefs must be aware of.

The Art of the Lecture

Marie-Odile Monchicourt: If a culinary revolution is needed, how can it be instigated?

Hervé This: The most important thing, I believe, is to explain. Lectures that simply rehearse the details of esoteric chemical reactions may well convince the audience that it is listening to a learned person, but the cause of cooking will not be advanced. To bring about a culinary revolution, lectures are needed to make the public aware that science is a wonderful and simple thing; also seminars, articles, and general-interest books. In promoting all these activities one is well advised to consider what the great scientists of the past have done. The example of the English chemist and physicist Michael Faraday (1791–1867) is wonderfully suggestive in this regard.

M.-O. M.: Why Faraday in particular?

H. T.: First of all because, despite a wretched childhood, he became one of the greatest scientists of his time—through sheer willpower, a determination to improve his mind, a great capacity for work, painstaking experimentation, careful reflection, and a commitment to lecturing and teaching. Faraday started out as an errand boy, delivering newspapers for Georges Ribeau, a French Huguenot bookseller and publisher who had immigrated to England. Thanks to Ribeau he learned to read and write; in particular, he made the discovery of *Mrs. Marcet's Conversations on Chemistry*, a popular introduction to the subject. This inspired him to devote what little money he had to learning about chemical experimentation firsthand. Gradually his abilities came to be recognized, and one day he was given tickets to a course of lectures by Humphry Davy, the great (and, moreover, worldly) chemist of the age who held a professorship at the Royal Institution of Great Britain in addition to being a fellow of the Royal Society, of which he was later elected president.

Faraday made the most of this unexpected opportunity, taking notes of the lectures and then making a fair copy, which he had bound and sent to Davy with a letter offering his services. Davy declined, but when his laboratory assistant fell ill a short while later, he remembered Faraday and hired him. Thus Faraday came to assist Davy in his experiments at the Royal Institution, and later to accompany him on a scientific tour of Europe. Faraday went on to become one of the great scholars of his day. It is to him that we owe our first experimental knowledge of many electrical and optical effects, including the discovery of benzene, the fundamental laws of electrolysis, and the liquefaction of chlorine and other gases.

M.-O. M.: And what does this have to do with the popularization of science?

H. T.: Faraday had succeeded Davy as professor of chemistry at the Royal Institution, which was comprised of several laboratories. The Institution was on the verge of financial ruin, and Faraday cast about for a way to save it. The idea came to him of organizing a subscription series of Friday Evening Discourses, as they were called, which continues still today in London. The industrialists who attended these lectures, hoping to develop technological applications for the scientific results described by Faraday, helped rescue the Royal Institution from bankruptcy.

The success of Faraday's initiative was the consequence in part of his having given careful thought to the question of what constitutes a good lecture, which he decided must include experiments. When you think about it, the idea makes perfect sense. We are not disembodied brains, unconnected to our senses; it is with the aid of our senses, after all, that we perceive the world. Experimental science is the hand, as it were, but it is not the hand alone that makes the gesture; the head, which conceives of the gesture, is indispensable to scientific discovery as well.

This image recalls the teaching of a certain school of Eastern philosophy, transmitted by the Chinese monk Shitao (1642–1707), known also by his monastic name Friar Bitter-Melon, who wrote *A Treatise on the Philosophy of Painting*. "If need be," he wrote, "one can paint without brush, ink, or paper. For the gesture has to have been conceived in such a way that, when one makes it, it was already there." Similarly, scientific experimentation has two related virtues: On the one hand, it shows in a concrete way what one seeks to demonstrate, or, at least, gives a tangible hint of it; and, on the other, it encourages us to seek theoretical explanations, which underlie and guide experiment.

Performing experiments during a lecture is also a way of overcoming shyness. When the lecturer does an experiment, this is what people in the audience look at, what concentrates their attention—not the lecturer himself, who is hidden behind the experiment, as it were. Faraday went on to give another important series of lectures, this time for children. These talks—the Christmas Lectures—were once again filled with experiments, and from them he created a wonderful, popular book, *The Chemical History of a Candle*, that recounts a group of experiments based on a simple candle.

M.O.M.: Are experiments the only reason for the success of your own lectures?

H.T.: Claiming any great success for myself would be presumptuous; but it is true that, as far as possible, I try to make use of the ideas of all the great teachers of the past. As the Greek playwright Aristophanes said, "Teaching is not about filling up jugs, but about lighting a fire." The French physicist and astronomer François Arago (1786–1853) had a rather different view, but a great deal of success—although it must be said his secret was shameful. He is said to have looked out at his audience, searching for the person who appeared the stupidest and then speaking directly to him, on the theory that if the stupidest person in the room understood, the others would understand as well. My father, by contrast, advised me never to underestimate the intelligence of the slower students in my classes.

Above all, I have noticed that if the lecturer gives the impression of being bored, there is little chance that the audience will be entertained. You must look as though you're having a good time. You can't just stand there. You've got to move around, tell stories, anecdotes—even walk up the wall if that's what it takes to make your listeners share the fun of learning. Some of my colleagues, either out of grumpiness or jealousy, complain that I'm a show-off, but they miss the point. It's entirely a question of communicating effectively. I could give boring lectures—but how then could I hope to arouse enthusiasm for knowledge?

6

A New Kind of Chocolate Mousse

Or how to make a chocolate mousse without an egg. For this final dessert, let's take as our point of departure the essential idea that cooking is about love, art, and technique.

Making others happy is easier said than done, but for the sake of the future of cooking we must learn to do it. As for art, some challenge the idea that it can exist in cooking at all, because they have not understood what it involves. It will always be possible to feed one person, or ten, or a hundred, or a thousand by repeating the same old dishes: cassoulet, choucroute, baekeoffe, bourride, fougasse—or for that matter a grilled sirloin steak with French fries, a cheese omelet, osso buco, and so on. The cook who reproduces a classic dish of this sort, possibly modifying it slightly, is an artisan, not an artist.

The artisanal cook, or craftsman, stands in contrast to what might be called "inspired" cooks, whose ambition is to dream up dishes that don't exist, to use ingredients to make works of art that will express a sentiment, arouse an emotion. Their objective is not to fill up the stomachs of their guests, but to produce culinary art. Why, then, does one speak of the "culinary arts," in the plural? Does anyone refer to the "musical arts"? No—there is only one musical art, just as there is only one culinary art.

Is the purpose of the artist to be "creative"? This is a fashionable word, popularized by a certain snobbish style of advertising, but the question remains: What constitutes creativity in cooking? Surely it does not involve repetition—the hallmark of artisanal cooking. Most of the facile innovations one encounters today emanate from food science. By adding sodium alginate (a gelling agent extracted from algae) to a flavored liquid (such as a bisque or a spicy sauce), for example, and then dripping this solution drop by drop into a bowl containing water enriched with calcium salts, you wind up with little gelatinous balls that have the flavor of the liquid in which the alginate has been dissolved. Nothing could be simpler—this is how the food industry has long made artificial caviar.

The existence of such a product is hardly a proof of culinary creativity. Indeed, it's not even worth talking about—still less since for some years now molecular gastronomy, particularly through the medium of the INRA seminars, has sought to devise technical innovations that will stimulate cooks to invent new dishes. For example, by melting chocolate in water and then whisking this mixture in a chilled bowl over ice, you get a true chocolate mousse (not merely one *au chocolat,* with some chocolate in it). This innovation, which I have named Chantilly chocolate, is a new technical resource—the equivalent of a new color of paint created by science for painters. Nonetheless Chantilly chocolate is not by itself a new dish, because a dish is a composition whose various elements must be modified and integrated with one another. Like fake caviar, Chantilly chocolate is little more than a gadget if cooks do nothing further with it. It is missing art and love. To confuse the development of new techniques with creativity is therefore to commit a profound error. It is also proof that one has not understood the essential fact about cooking, namely that flavor is what matters most of all.

From this it follows that culinary art does not consist in making dishes that are pretty to look at. A little sprig of parsley on the side, a drop of sauce in the form of a comma in a corner of the plate, a pleasing arrangement of ingredients selected with an eye to shape and color—these are merely superficial touches. Cooking, I repeat, is first and foremost a matter of flavor. The visual aspect contributes to the overall effect of the dish, of course, and it is certainly a good thing to take care with the presentation of the dish. But in the mouth it is the appreciation of smells, tastes, and texture that counts. Culinary creativity is mainly concerned with achieving a harmonious relationship between these three elements, on the one hand, and the shape and color of the ingredients of a dish, on the other. Whoever forgets this understands nothing about cooking. Spinning and pulling sugar has nothing to do with cuisine; it is a minor branch of sculpture.

Chantilly Chocolate

Marie-Odile Monchicourt: You invented Chantilly chocolate in 1995. How exactly is it made?

Hervé This: What I like about this invention is that it was the result of a generalization, on the one hand, and that it provides a stimulus for further innovation, on the other.

Usually you make a chocolate mousse by melting chocolate, along with butter and egg yolks in order to enrich the flavor, and then folding the mixture in a mousse. This gives you a mousse *with* chocolate. Chantilly chocolate is a mousse *of* chocolate.

M.-O. M.: So where did the idea come from?

H. T.: From thinking—particularly about whipped cream. Let's begin with milk, which can very be roughly described as a dispersion of fatty droplets in water. Milk is an "emulsion"—from the Latin word *emulgere*, meaning "to milk." When milk is left to settle, the fatty matter rises to the surface, forming cream. This is also an emulsion, but one that has a greater concentration of fats than milk. It is not uninteresting, by the way, to observe that cream is whiter than milk, and whole milk whiter than skimmed milk. The fewer fatty droplets dispersed in the water, the less reflection there is of the light—typically white light—that illuminates the liquid.

To make whipped cream, you beat the cream with a whisk, which introduces air bubbles into the emulsion; these air bubbles are gradually trapped by the fatty droplets, which in the meantime have fused around them. When sugar is added to the whipped cream it is called Chantilly cream.

To make Chantilly chocolate, you follow the same procedure: You take an emulsion and fill it with air bubbles. Nonetheless the ingredients are different. The emulsion is obtained by heating chocolate in water. On melting, the chocolate releases fatty droplets, which become coated by the lecithin molecules that were added to the chocolate when it was made. In other words, whereas adding water to chocolate causes the chocolate to gain in mass, adding chocolate to water produces a chocolate emulsion. Then, when you whisk this

This mille-feuille created by the Paris chef Christéle Gendre exploits two applications of molecular gastronomy: roasted flour (for the puff pastry) and Chantilly cheese, made according to the same principle as Chantilly chocolate.

emulsion, while keeping it chilled, you end up with a true chocolate mousse: Chantilly chocolate.

M.-O. M.: Are there certain proportions that must be respected?

H. T.: Of course. There must be enough fat to trap the bubbles. You will get good results using 225 grams (a bit less than 8 oz) of dark chocolate for 200 grams (7 oz) of water. The consistency will then be that of Chantilly cream. And if the preparation fails for one reason or another, you can rescue it by keeping in mind that success depends on achieving the right balance between the constituent molecules. Whether there is too little fatty matter or too much, you can redress the balance by adding what is missing. If, for example, you heat the chocolate in a liquid and, after whisking, the result remains liquid, the reason must be that the quantity of fat—which is to say chocolate—was insufficient. Put the pan back on the burner with some more chocolate, then whisk the emulsion again over ice cubes. Conversely, if your mousse is too firm, this means there isn't enough water (or other flavored liquid). In that case, melt it over low heat, add a little liquid, and whisk again over ice.

M.-O. M.: Are you proud of your invention?

H. T.: There's nothing very special about Chantilly chocolate in a way. After all, it's so simple—and in any case it's only one of an infinite number of possible new preparations, since the principle of Chantilly is not limited to cream and chocolate. You can use any kind of fatty matter that solidifies at room temperature: foie gras, butter, brown butter, cheese, and so on. Further variations of flavor are possible as well, depending on the liquid that is used. For Chantilly chocolate, for example, this liquid can be any aqueous solution: orange juice, tea, coffee, bouillon, wine, and so on.

I'm not sure how much pride one can take in such an invention. The whole thing is very simple, as I say. And once you succeed in generalizing a certain culinary principle, applying it over the range of suitable ingredients means that invention becomes mechanical. Innovation, creativity, invention, discovery— lurking behind all these words is a bit of fantasy. Should we really be interested in satisfying our ego, or should we devote ourselves instead to discovering what fascinates us? Shouldn't we be proud only of what we will produce tomorrow?

Novelty: Between Desire and Fear

It is a cause for some surprise, as we have already noted in passing, that human beings should to one degree or another exhibit a taste for novelty in their diet. Why aren't we content with a single dish—a choucroute for those of us from Alsace, a cassoulet for those from the southwest of France, pancakes for those from Brittany? After all, we never grow tired of eating these regional specialties; indeed, we always want them to be prepared in the same way, and tolerate only those variations that everyone agrees improve the dish.

Repetition of this sort perfectly suits the primate in us, by reinforcing the neophobic instinct that protects us against the dietary dangers of the natural world. For every edible plant, how many others are poisonous? Plants are flavorful because they contain substantial concentrations of sapid and odorant mole-

This green apple Chaptal (a concoction I have named after the French chemist Jean-Antoine Chaptal [1756–1832]) is made by dissolving sugar and gelatin in green apple juice over heat and then whisking this mixture. The result is a very large volume of mousse, because good use has been made of the foaming properties of gelatin. A simple calculation shows that several quarts of floating island can be made in the same way from the white of a single egg.

cules, many of which are harmful to human beings. I have already mentioned the maxim of the Swiss physician Paracelsus (1493–1541), who held that it is the dose, or proportion, that determines whether something is safe to eat. In other words, everything is poisonous, but in sufficiently small doses the risk becomes negligible. Even so, we wouldn't be here talking about human dietary preferences if our primate ancestors had eaten anything they wanted—and we won't be here much longer if chefs entertain themselves by seasoning our food with whichever herbs catch their fancy. Nature—let me stress the point again—is bad for you. Lightning is natural; the wolf is natural; so are cholera, flood, mudslides, volcanic eruptions—and death, which eventually strikes down all of us.

But if food neophobia has served us so well for so long, why do we have so great an appetite for novelty? Why *aren't* we content with the same dish day after day? Because as human beings we have an equally innate urge to explore the world around us. We are continually trying to go farther, to break free of the boundaries of our culture—even of our planet. We are permanently torn between our taste for the familiar and our taste for discovery. It comes as no surprise, then, that chefs, who are obliged to try to satisfy both tendencies in the domain of food, should be in the habit of solidly grounding themselves in the traditional, the known, while permitting themselves small departures here and there. A choucroute, yes—but with a little pine honey added to it. Or even a choucroute cooked in lime blossom tea.

What Will We Eat Tomorrow?

I hope that tomorrow we will eat what we decide to eat—and that some people will have had the daring to propose new ideas, and the courage to defend them. However this may be, it is clear that the French chemist Marcellin Berthelot (1827–1907) was badly mistaken a century ago when he predicted that in the year 2000 we would be eating food tablets. Berthelot had miscalculated the amount of energy necessary for a human being to survive. Even if one were to eat pure fat—the most energy-filled substance there is—one would still have to consume at least 300 grams (10 oz) a day, or more than two sticks of butter. People who fear a diet of synthetic pills can put their minds at rest.

But if we're going to eat 300 grams of fat, why don't we choose foie gras rather than a neutral fat, and serve it on top of a piece of good crispy bread? We'll get more out of it that way, especially because evolution has equipped us with a complete sensory apparatus for appreciating food that is connected to pleasure centers in the brain. (Berthelot made another and even greater error in failing to take this into account.) The contrast between the crispiness of the bread and the smoothness of the foie gras, as well as differences of flavor, smell, and temperature, are all necessary if we are to be happy consuming our minimum daily energy requirement in this fashion.

What will we eat tomorrow? Without a crystal ball it's very hard to say. Cautiously, we may venture to predict that tomorrow we will eat what we invent today.

RECENT DEVELOPMENTS

Whether traditional-minded food writers like it or not, molecular cuisine will be part of tomorrow's culinary landscape, for the simple reason that it has already succeeded in creating a place for itself—and because it is fashionable, just as nouvelle cuisine was in the 1970s. Even young cooks who have been led to believe that the only good cooking is classical French cooking can no longer ignore the fact that the classical tradition now has competition. They have become aware, not only from their culinary training but also from the cultural environment in which they live and work, of things like liquid nitrogen, which makes superb ice creams and sherbets in a few seconds; various gelling agents extracted from algae, which in recent years have supplanted gelatin; magnetic induction cooktops, which now rival the performance of gas burners and electric coils; Chantilly chocolate and all the other new ideas that Pierre Gagnaire and I post online every month on Pierre's Web site, free of charge, so that these innovations will be widely known and used. And we are not alone. There are many other cooks who are no less committed to innovation—proof that the conservative cause is already lost.

Indeed, it is not going too far to say that classical cuisine has now been superseded, just as the cooking of Guillaume Tirel had been superseded by the end of the Middle Ages, and that of Carême by the end of the nineteenth century. It's

This dish by Pierre Gagnaire, known as a "Buren" (after the French conceptual artist Daniel Buren), is an example of what might be called abstract, or nonfigurative, cuisine. This involves giving gustatory pleasure without relying on the familiar cues provided by the shapes, colors, flavors, and smells of classical dishes.

too late to regret that the past is past. We will no more go back to roasting meat over a wood fire (for purposes of everyday cooking, at any rate) than we will to lighting our homes with candles, and even though some home cooks still make quenelles using a sieve, nowadays in restaurants only food processors are used to finely mince the meat (partly because they save time and effort, but also because they do the job better).

And since progress is on the march (it always has been, by the way), aren't we better off trying to guide the course of events in a direction that conforms more closely to our fundamental values? In the case of cooking, health is an important consideration, along with safety, flavor, and—if I may say so once again—the happiness that we owe to those whom we feed.

Health, Food Safety, and Flavor

First of all, there is the question of health. At a time when a pandemic of obesity rages, when even on the Greek island of Crete (famed for its diet, whose emphasis on fruits, vegetables, whole grains, and olive oil is thought to be one of the most healthful in the world) more than a third of the young people are overweight or obese, governments are now beginning to take action, launching information campaigns aimed at getting youngsters to eat more fruits and vegetables and to exercise more. Such campaigns cannot be expected to have an immediate effect, but they are essential if we are to prevent our children from dying in their fifties from cardiovascular diseases. Do chefs have something to contribute to this effort? I believe they do. Much can be done, for example, to improve the cooking of fresh vegetables. Enough of steaming and boiling. We need to find better ways to pique people's interest.

Safety must not be neglected either, of course, though it is unlikely that regulation will be relaxed. With each new food scare some government officials lose their jobs, and so their successors hasten to take steps to avoid the same fate. Contamination by microorganisms is the main concern of the health authorities, and rightly so. One may wonder, however, whether enough attention is paid to the dangers of toxic molecules such as estragole, present in tarragon, or trimyristicin, which is found in nutmeg. Are the so-called edible orchids that are sold in markets today perfectly edible? Many food products present some degree of health risk that needs to be investigated more systematically.

Finally, we must recognize that the importance of flavor is directly related to the question of health. The science of nutrition has amply demonstrated that one eats less if one learns to eat more slowly and if one eats well. To eat well is to eat foods of varying texture and temperature with contrasting tastes and smells. Because the perception of flavor involves all of these things, it is reasonable to suppose that a better understanding of it, communicated by new methods of teaching that place emphasis on the latest scientific research, will help young chefs to raise standards of health by creating more flavorful dishes.

Looking Ahead

I have already mentioned the fashionableness of molecular cuisine, recently highlighted by a ranking of the world's top chefs (an idiotic idea, I quite agree) in the British magazine *Restaurants*: first place, the Spanish chef Ferran Adrià; second, the British chef Heston Blumenthal; third, the French chef Pierre Gagnaire. All three apply ideas from molecular gastronomy in their cooking.

I must take this occasion to correct another misapprehension. If Pierre Gagnaire consents to lend his name to the cause, he does so out of friendship toward me, and without sacrificing on the altar of technology the artistic aspect of his cooking. On the one hand, he has absolutely no need of the innovations that I propose every month (just as Rembrandt could have drawn with only a charcoal pencil if he had wished); and on the other, he declines to make use of many of the ideas that I suggest because they do not suit his artistic vision. Pierre does not do molecular cuisine; he makes culinary art, in his own style.

In any case, molecular cuisine (not, let me emphasize once again, to be confused with molecular gastronomy) is already dead, since it is fashionable. Today's innovators will soon abandon it in order to explore new territory. Note-by-note cooking, which makes use of pure molecules (tartaric acid, glucose, polyphenols, and so on), has begun to be practiced in restaurants throughout the world, but I have come to believe in the meantime that culinary constructivism, which I mentioned earlier, holds still greater promise—since chefs are now finally in a position to predict the gustatory effects of their compositions.

More generally, these different tendencies seek to communicate the idea that, in addition to the recipes of classical tradition (which codify a set of fixed procedures, slightly varying the flavors), there is now the possibility of completely constructing a dish, with due regard for all of its component elements. Obviously this is a much more difficult approach to cooking, but not a less flavorful one—to the contrary! We may look forward to the day when passionate young chefs, having mastered the full range of gustatory forms and combinations, will create brilliant new compositions that will take their place alongside the great triumphs of the past. There will then no longer be only classical cuisine, but classical cuisine enriched and extended by modern innovations.

Note-by-Note Cooking

Until now, cooks have made use of carrots, meats, fish, eggs, and so on. Consider the carrot. It is an assemblage of water, cellulose (think of absorbent cotton), pectins (which make jams set), some sugars (fructose, glucose, sucrose), amino acids and organic acids, and molecules that in very small concentrations give bouillon its characteristic smell, and therefore its flavor. All these molecules are simultaneously deposited in the pan at the very moment when the carrot is thrown in—as if a pianist could play only chords. What heavy music!

It is true that chefs have encouraged farmers to diversify the number of varieties in order to obtain different nuances of flavor, but the problem remains: A carrot, like a chord, is still a lumbering creature. There is an alternative to this "natural" enterprise—which in reality is highly artificial, since new varieties have for the most part been deliberately introduced by crossbreeding, with the fortunate result that the carrots we know today do not at all resemble the thin, fibrous wild specimens that our ancestors foraged. The "synthetic" alternative is to add sodium chloride (better known as salt) or sucrose (wrongly called sugar—as though there were only one kind) to carrots as they cook. This amounts to adding a note to a gustatory chord, so that we can hear the flavor more clearly, as it were.

Beyond this, we can add organic acids (tartaric acid, citric acid), amino acids (glutamic acid, alanine, tyrosine), or phenolic compounds (such as the ones that give color to wine). Will the cooking of tomorrow be synthetic, or note-by-note as I call it? The possibilities for creating new and more precise flavors would be greatly increased in that case, of course, but perils await explorers, and the job of clearing a path through the culinary wilderness will not be easy. On ancient maps, beyond known frontiers, one finds the words *Terra incognita, hic sunt leones* (Unknown land, here be lions), indicating that danger lay ahead. Undoubtedly there are many dangers in culinary exploration as well. Happily, however, cautious and reasoned exploration is also possible.

Chemistry, I maintain, is no more responsible for gas warfare than the Curies were for the atomic bomb. The people who make the bombs and the lethal gases, as well as the people who actually use these terrible weapons, are the ones who bear responsibility for the deaths they cause. As a chemist, which is to say as a scientist, I reject the notion that molecular gastronomy should be held responsible for the ways in which molecular cuisine is being developed today throughout the world. I do, however, claim responsibility and credit for some of the advances in knowledge on which these novel applications are based, while urging that such experimentation be subject to critical analysis and open debate so that tomorrow we do not suffer from the unintended consequences of innovations that were not sufficiently thought through beforehand. Note-by-note cooking, to take only one example, is an immense unexplored territory. Clearing the way forward will not be easy, and the Christopher Columbuses of this new continent will have to withstand torrents of gastronomical criticism, sometimes running aground on the distant shores of the inedible. There is nonetheless a new world to be opened up, and they must have the courage to try.

The fact of the matter, of course, is that note-by-note cooking is already here. In France alone there are cooks who haven't waited for this book to come out. Having heard one of my lectures, they have begun to try out new ideas on their own. Pierre Gagnaire has been experimenting with sauces that I have named "Wöhler sauces," in honor of the German chemist Friedrich Wöhler (1800–1882), who synthesized the first organic molecule. Others have begun to investigate the possibilities of note-by-note cooking as well, because it is simple (one has only to pour a very small amount of an ingredient in the pan, as one would add salt to a dish), inexpensive (the necessary chemical agents are sold by the ton), and requires chefs to do nothing they aren't already used to doing (like adding salt to a dish). Tomorrow's trend has therefore already gotten started today.

Reforming Education

Culinary instruction will in any case have to change. If courses of nutritional education are introduced in primary and secondary schools, not only can we begin to combat the pandemic of obesity, but kids who choose a cooking career will start out with a basic technical knowledge that culinary schools will have to take into account in designing their programs. And if new ideas about culinary art and the love that chefs owe their guests come to be widely shared, our whole way of thinking about food will have been changed for the better as well. The way we cook tomorrow will depend on what we are willing to do today.

A NEW PROFESSION: CULINARY ENGINEERING

A great many false ideas—for example, that cooking proceeds by concentration (it was long claimed that roasting a piece of meat causes its juices to seek refuge from the heat in the center) or by expansion (meat was thought to dilate when cooked in water)—are now being eradicated the world over as molecular gastronomy develops and disseminates new knowledge. The disappearance of such misconceptions will free chefs to look at classical dishes with a fresh eye while inventing new dishes. The development of molecular gastronomy may also be expected to give rise to a new profession: culinary engineering. Since science produces only knowledge, it is up to engineers to apply the results obtained through basic research. The profession already exists in England and Spain, where Heston Blumenthal and Ferran Adrià employ technical specialists.

In France, we can rejoice at the recent creation of the Foundation Food Science and Culture Fondation under the auspices of the Academy of Sciences. The organization is divided into a series of divisions whose work is conducted through a network of regional programs. There are six divisions, each with its own administrative staff: science, technology, art, hygiene/safety/regulation, training, and communication. The regional programs are meant to coordinate research in different parts of the country in order to optimize each region's contribution.

A CULINARY BRAIN

The purpose of this new foundation is to provide the culinary profession in France with the technical center it has lacked until now. Alongside the INRA seminars in molecular gastronomy, university courses in molecular gastronomy, molecular gastronomy workshops in cooking schools, and laboratory workshops

In this dish, also by Pierre Gagnaire, an egg yolk cooked to 67°C (153°F) is set on a bed of Chinese noodles cooked with cream and the white of the same egg, heated and then whisked.

on flavor in schools, it is now possible to organize a series of research projects aimed at promoting a scientifically informed approach to cooking.

The effectiveness of the foundation's various initiatives will be in proportion to the number of cooks who take part. If enough people join in, and if neither world war, nor a pandemic of avian flu, nor global energy shortages intervene, we can be sure of enjoying tomorrow what we begin constructing today.

Culinary Constructivism

Marie-Odile Monchicourt: On several occasions you have mentioned culinary constructivism. What does it involve?

Hervé This: It involves a new way of cooking. I chose this name by way of contrast with "deconstruction," the spirit of which I don't like. Think of a pot-au-feu: Classically it contains meat, bouillon, and vegetables. Deconstruction in this case consists of serving the meat in powdered form, for example, the bouillon as a jelly, and the vegetables as a sherbet; or the meat as a jelly, the bouillon as a powder, and the vegetables in razor-thin slices; or the meat as a powder, the bouillon as an emulsion, and the vegetables in julienned strips; or—well, you get the idea.

There are a great many possibilities, as you can see; the application of new techniques enlarges the number still further. Which one should we to choose? None of them has any real justification, and this is why I don't like this way of cooking. Like music, cooking is a cultural enterprise, and a plate of food must have meaning, for otherwise it risks becoming the equivalent of the noise made by a monkey striking the keys of a piano at random.

I recently ate a celery sherbet (very well concocted, by the way) with powdered peanut butter. First one sensed the coolness of the celery, then the slightly disgusting fattiness of the peanut butter. Why did the chef compose the dish so that it would be experienced in this way? Because he didn't understand that the classic dish from which his creation derived (a favorite of American teenagers, who spread peanut butter on stalks of celery) is intelligently assembled. In eating a dish, you sense first of all what is on top, and you finish by sensing what is beneath it; so, in this case, the classical version leaves your mouth with a sensation of coolness at the end, due to the celery, which you encounter after the peanut butter.

I don't like deconstruction in cooking because, ultimately, it represents a form of laziness on the part of the cook. We should construct instead. How? Culinary constructivism, in the form that I originally proposed, sought to make people cry with emotion, or laugh, or feel anger, just as the other arts do: music, painting, cinema, photography, literature—

M.-O. M.: That isn't within the reach of every cook.

H. T.: You're right—and this is precisely why I then tried to recast culinary constructivism in another form, which can be thought of as a branch of the first. This new version makes use of inventions that I call *checkerboards* (two- and three-dimensional compositions of various ingredients), *fibrés* (artificial kinds of meat and fish), and *conglomèles* (artificial fruits and vegetables). These artificial objects are in the grand tradition of French culinary invention. It was Carême, after all, who introduced architecture into pastry-making and who, more generally, was the champion of what might be called monumental cuisine—a constructive style if ever there was one.

In Carême's time, construction was limited to reproductions of Greek temples, Chinese houses, palaces, and the like—pretty, but devoid of culinary meaning, just like the sculptures in sugar or ice or butter that sometimes clutter up banquet presentations today. I shall never forget visiting Cuba, where cooks lack basic ingredients, and seeing a sculpture more than six feet high made of butter, which began to melt from the heat. This sort of thing isn't a part of cooking; it's a form of sculpture, and the authors of such works ought to be compared to Rodin rather than to Édouard Nignon or to Carême (next to whom they cut a rather sorry figure on the whole). What is cooked must be eaten—otherwise it's not cooking.

To come back to Carême and his ambition to introduce architecture into cooking. The idea is by no means a bad one, much less outmoded; but today we need something other than palaces, pagodas, and temples. I propose that we construct monuments of and for our time: counterparts to Beaubourg in Paris, for example, to the arches at La Défense or to the Institute of the Arab World along the Seine. Naturally I'm not suggesting that chefs should create dishes that resemble these buildings, because cooking isn't architecture. Instead I propose—and this is what the new constructivism is really all about—that dishes should at long last actually be constructed, instead of simply being arranged on a plate.

At Daniel Boulud's restaurant in New York, for example, I was served a shallow bowl containing small potato *échaudés* (he called them gnocchi, but I prefer the classical French term because etymologically *échaudé*, from the verb meaning "to scald," is more accurate), a brunoise of cèpes, or porcini mushrooms (about three-eighths of an inch per side—what an intelligent size!), and a julienne of black truffle (not too thinly sliced)—all of these ingredients bound together with a very thick, highly gelatinous sauce to which a touch of butter had been added. The contrast of textures in this dish was remarkable. But imagine if the elements, instead of being arranged in a shallow bowl, had been assembled vertically in layers (or checkerboards, as I call them). In that case the contrast between the softness of the *échaudés* and the hardness of the cèpes would have been still more pronounced. This is the sort of constructivist cuisine that I am impatiently waiting for!

Constructing a Menu

In the preceding pages I have several times mentioned my friend Pierre Gagnaire, and there is a risk the reader will suppose that compliments I have showered upon him are undeserved or that they are sent his way only in the expectation of similar praise in return (pass me the rhubarb, I'll pass you the senna). This isn't the case. One will have a better idea of Pierre's extreme attention to detail when one knows how he works—not only in constructing individual dishes, which we have discussed at length in our book *Cooking: The Quintessential Art,* but in constructing a complete meal.

Let's go back a few years. Pierre and I were coming back from a joint lecture overseas. The airline announced that the plane would be delayed for four hours. What to do? Work, of course.

Pierre took out of his briefcase a copy of his new menu, which had to be finalized the next day. A galley had already been printed on fine paper. Pencil in hand, he began to reread the text in order to be sure that everything was in order. I peered over his shoulder. "Why a veal pâté at the beginning of the menu, followed by a partridge?" I asked. Pierre thought for a moment, and then tore up the page. "You're right, it doesn't work."

I was right? All I did was ask a question, the abrupt answer to which I didn't understand. He then took out a pad from his briefcase and began to compose a new menu. Halfway through, he tore up the page, and on the next one began once again to write down a list of dishes. He tore up this page as well, and started on another, and so on, and on—indefatigably.

I didn't want to disturb him too much. But I was curious to learn the reason for these hesitations. What displeased him about these menus? The overall effect, perhaps? No. The flavors, because they didn't build in intensity from dish to dish? Not that either. Something about the tone of the composition? Nor that.

Four hours went by, and finally the plane took off. We arrived back in Paris and said goodbye. "We got a lot done, didn't we?" Indeed we had—but the menu still wasn't decided.

The next afternoon Pierre telephoned to say that he still wasn't happy with what he had come up with. He had spent the whole day in his office, but it was no use. The day after that, same thing. The days went by, and the new menu *still* wasn't ready. Not one pan had been touched for this new menu. A wastebasket filled with crumpled paper was the only sign of activity. Then one day my phone rang. "That's it, I've got it! I'm faxing it to you."

The page came through that evening, but the next morning I was told to throw it out. And it was only much later that the menu was finally settled.

In retrospect, Pierre's reasons for being dissatisfied are clear enough. If cooking is first and foremost an artistic activity, as it is for Pierre and some other chefs, then style is important. It is not only a matter of getting the technical aspects of a composition right. In this case, of course, the "melodic" line was right; naturally, the dishes were placed in the correct sequence in each of the various drafts—all of this was beside the point. The point, metaphorically, is that you can't have an element of Picasso's blue period in a picture from his pink period. And for a prolifically creative artist, periods succeed one another very quickly.

This tale teaches the rest of us modesty. What amateur cook insists upon such a degree of perfectionism? We generally content ourselves with taking a reasonable amount of care in assembling a meal, so that the various dishes harmonize with one another and form a coherent menu. But none of us goes this far.

A CUISINE FOR MODEST AMATEURS

The story of Pierre and his menu shows that amateur cooks can't compete with the great professionals. Why should they even try? Jorge Luis Borges (1899–1986) distinguished between black envy, which leads us to destroy what we do not have, and white envy, which kindles a desire to improve our abilities, out of admiration for the talents of others. This story inevitably cuts those of us who like to think of ourselves as cooks down to size, but it can also help focus our efforts and inspire us to do better. As far as the technical aspect of cooking is concerned, we are all capable of achieving a satisfactory degree of competence, with a little reflection and the sort of scientific knowledge that molecular gastronomy provides. As for the artistic dimension, it is only by studying the work of the great artists that we will progress farther. And as for love? Here, there is no alternative to looking inside ourselves, in order to find ways to give happiness to those whom we feed. What a fine task for us to undertake!

The Unending Conversation

DEAR FRIENDS,

Looking back, it appears to me that these interviews with Marie-Odile Monchicourt have often led me, shamefully, to indulge my ego. For a person's ideas are of no consequence if they are not shared, in which case the "I" is indecent.

The ego is a hateful beast. Obviously there are many excuses. First, this book grew out of a series of interviews. In an interview, one opens up (as Brillat-Savarin rightly said, one must speak without pretension); one gives in to the curiosity of the interviewer, agrees to her request that one share small secrets. The trivial becomes a form of politeness, and confidence develops as a kind of conviviality.

On the other hand, Marie-Odile Monchicourt feels that science is more accessible if the scientist steps out of character (or at least out of a certain stereotype), if he agrees to show his human side, to show that that he lives like everyone else—which is true. She also feels, I think, that learning about the great scientists of the distant or recent past helps bring their achievements alive, that a living science is more interesting than a dead one. And this is true as well: Reading biographies of the great scientists excites the interest of young people who are considering a career in science, just as reading accounts of the great chefs arouses in others a desire to follow in their illustrious footsteps. It is for this reason, in the preceding pages, that I have mentioned Lavoisier, Chevreul, Faraday, Carême, and so on. Not that I dare to compare myself to them, of course, but I do not think it indecent to make a point of mentioning my great admiration for these remarkable figures; nor do I think it unuseful to try to communicate the enthusiasm I feel for their magnificent discoveries.

This raises an important distinction between what one is (a chemist or a chef or whatever—which isn't very interesting) and what one does. That's the thing. What one does, what one is about to do (not some vague project for the future, but something that's getting under way right now), news of which deserves to be shared—not in order to provoke undeserved admiration, but because the sharing of intellectual and human adventures brings with it the possibility of making new friends.

There is therefore unavoidably some amount of ego in the preceding pages, but it is meant above all as an invitation to join me in exploring the essential problem of cooking: how to give greater happiness to others. Once the question has been posed, it becomes more and more insistent—and more and more challenging. Answering it will require the intelligence, care, attention, and effort of an immense number of people passionately devoted to cooking.

At all events, if these excuses persuade you to volunteer your services for this difficult but immensely rewarding enterprise, so much the better!

Artisans and Artists

The preceding pages are an argument for promoting knowledge, without regard for national or disciplinary boundaries, by bringing together chemistry, physics, and biology with cooking, architecture, painting, music, and philosophy. For in treating cuisine as a manifestation of culture, we need to draw upon the methods of all these fields in order to clear the way for further cultural exploration.

Readers, my friends, don't hesitate to join in—and by your own singular abilities contribute to this cultural banquet. Cast your gaze upon the culinary world, its customary practices, its usual assumptions, and communicate your insights to the community of those who wish to understand them and make use of them, and in this way help to improve an activity that amply deserves our assistance. How do you show love to your guests? Don't hesitate to come to the INRA seminars on molecular gastronomy when you are in Paris, or to the meetings of the various divisions of the Food Science and Culture Foundation, and share your experience.

Do you have the wisdom of the ancients? Use it to tame the wilder impulses of the most ardent moderns. Do you have the freshness of youth? Don't shrink from advancing new ideas, from refuting the comfortable assumptions of your elders, assumptions that you have not yet had time to metabolize. Are you skeptical? Feel free to express your doubts, in order to encourage the most assured among us to think better of caution. Are you aware of new facts? Share them with the community, so that tentative conclusions can be drawn from them, and then tested.

There is much to be done, on both a modest and a grand scale. Consider, for example, the writing of recipes. Recipes have often been denounced in the preceding pages as little more than degrading protocols. Indeed they are—but how, then, are we to transmit a culinary idea? Without giving weights and measures, without precise instructions, the amateur cook hesitates. But if the instructions are too precise, there is danger as well, because animal and plant tissues are variable in ways that the cook must take into account.

How, then, are we to envisage a new way of transmitting new culinary ideas, superior to the kind of description we are used to? How are we to impart technical intelligence, on the one hand, and artistic intelligence on the other, in a way that is both appetizing and digestible?

The grand question, obviously, has to do with love. I come back to it time and time again, as you see. We will have to dig deep, to work hard—and that's the wonderful thing, the happiness that lies in store for all of us.

The roar of war, tragedy, and catastrophe can be heard throughout the world today. But the experience of Lord Rumford, an eminent scientist and adventurer of the late eighteenth and early nineteenth century, illuminates our predicament. When the elector of Bavaria invited Rumford to Munich, he arrived to find a city full of beggars and thieves, and he sought a remedy. Instead of preaching morality, he set about feeding the people—and morality followed, as a result of eradicating hunger. May we not dream, however naively, of furthering the cause of peace and social justice by reintroducing in the schools an enlightened form of culinary instruction?

The essential thing, if fresh initiatives are to be productive, is that they make distinctions. One may be a good or bad artist, a good or bad craftsman. Let's forget the bad, keeping only the good—along with the distinction between artist and craftsman. What they do is not the same. Understanding this simple fact makes it possible to see the way forward more clearly. Similarly, science must be distinguished from technology. "What are you trying to do?" the French photographer Henri Cartier-Bresson (1908-2004) used to ask. "What does it involve?" Thus, too, cooking must be distinguished from gastronomy. Cooking produces dishes, whereas gastronomy produces knowledge. And so on. Doesn't knowledge consist in seeing clearly what one didn't see before?

That's it for now, time to say goodbye. Without nostalgia, because goodbye is also the promise of meeting again—at the great table of knowledge. Until we meet again, then.

Index

Page numbers in italics refer to illustrations and their captions.

Arts & Traditions of the Table

PERSPECTIVES ON CULINARY HISTORY

Albert Sonnenfeld, *Series Editor*